THE BUILDING ACTS AND REGULATIONS APPLIED

THE BUILDING ACTS AND REGULATIONS APPLIED

BUILDINGS FOR PUBLIC ASSEMBLY AND RESIDENTIAL USE

C M H BARRITT

LONGMAN

Addison Wesley Longman Limited
Edinburgh Gate, Harlow
Essex CM20 2JE, England
and Associated Companies throughout the world

© Addison Wesley Longman Limited 1997

First published 1997

British Library Cataloguing in Publication Data
A catalogue entry for this title is available from the British Library

ISBN 0-582-30201-3

Set by 3 in Baskerville 10/12 and Gill Sans
Produced through Longman Malaysia, PP

CONTENTS

PREFACE

This is the third in a series of books on the subject of the Building Regulations; its companions cover the Regulations as they apply to (a) Houses and Flats and (b) Shops, Offices and Factories. The aim of this volume is to examine all the other building types not dealt with in the other two. Clearly, this is no small task and, indeed, there must be many building uses not specifically mentioned in the text but to which the Regulations will apply. Generally, any building where people will assemble in number or any in which people reside, i.e. live and sleep, other than a house or a flat, would be subject to these provisions. A detailed list of typical building uses is given in Chapter 1.

The form of the text follows that in the other two books and endeavours to present the requirements of the Building Regulations in a more convenient form than is offered by the legislation itself. This is split into 14 separate Approved Documents each dealing with a particular subject such as 'Conservation of Fuel and Power' or 'Ventilation' and describing constructions that are accepted as meeting the requirements. This is quite logical as far as the legislators are concerned since each subject requires to be considered in detail by experts in the field. However, it means that each Document presents a lot of information that is not necessarily relevant to the project a designer or builder may have in hand. Furthermore, the Regulations applying to one particular element of the building are distributed through as many as ten different Documents, making it difficult to be sure that all aspects have been considered.

To this effect I have extracted the Regulations and approved constructions relevant to the building types covered in this text and regrouped them according to the element to which they relate. Thus, the chapter on walls, for instance, deals with all the Regulation requirements applicable to this element in an assembly building or one used for residential purposes.

There is a limit, however, to the extent of the coverage of the Approved Documents. With buildings of the wide variety and complexity covered in this range it would be impracticable to lay down standards for every structural and service provision employed. Instead, the legislators have relied on that vast reservoir of information, the British Standards, as well as on more specific sources of guidance such as the NHS Estates publications on standards in health care premises or the Home Office *Guide to Safety in Sports Grounds.*

These supporting documents are, mostly, advisory and recommend appropriate standards. They only become mandatory if the law makes them so – that is, if the Regulations state that to comply with the requirements the building must follow the standards laid down in, say, a particular British Standard.

To try to make this as complete as possible I have included a review of these supporting documents and the technicalities that I considered would be useful to the building designer or constructor. I could not include a full treatise on them all and have omitted those aspects that are only really of interest to the specialist. For instance BS 5268: Part 2 is a very comprehensive statement of the way to design structural timber frames. I have not attempted, in Chapter 4, to go into the analysis of the stresses to be met in a frame, nor the way to compute the sizes of the members. This would represent a complete book on its own. I have, instead, given an explanation of the way it is done, the principles involved and practical matters such as possible joints, the spacing of bolt holes, the use of timber connectors, etc., as this is, I believe, the information the designer, builder or site supervisor will find to be of the most benefit. Incidentally, on this subject I have also introduced a review of the changes coming with the introduction of the new Eurocode on the subject.

Other Eurocodes are in the course of preparation or production; some are out for consultation and all will soon be coming into effect, replacing the corresponding British Standard. Where relevant I have included such information as is available and within the scope of this book, even though it is, not yet, the standard that must be observed.

I have also included a review of the Construction (Design and Management) Regulations (CONDAM). Although these are neither part of the Building Regulations nor referred to in any Approved Document, they are of considerable importance to everyone involved with the construction of a building, including the client, and cannot be ignored in any book on building. The CONDAM Regulations place a statutory duty on all concerned to see that the construction work is planned and executed with due regard for health and safety on the site.

While the building function dictates many of the ways a building is designed or constructed to accommodate the particular needs of that use, not every aspect of the structure is thus controlled. For instance, whatever the purpose of the building, the foundation design will be determined solely by the nature and magnitude of the loads to be transferred to the ground. For this reason, many similarities will be observed between this book and the book on the Building Regulations applying to Shops, Offices and Factories. Rather than leave out any such duplication, I decided to repeat the information and thereby ensure that each volume is a complete treatise on its declared subject.

I believe that this book, like its companions, will be of great help to students trying to grasp the complexities of the current legislation surrounding the construction of buildings. I also hope that it will find its way onto the shelves of the libraries in design offices, builders' premises and even site offices where, as one reader has commented, it will provide answers 'on the run'.

C.M.H.B. *Colchester 1997*

ACKNOWLEDGEMENTS

To produce a book of this nature, many sources of information need to be tapped and I am grateful to all those people who responded so generously to my pleas for help. In particular, I would like to thank the staff at my local branch library in the Prettygate suburb of Colchester. Due to economies and 'rationalization', one of my main sources of information, the British Standards held in the main library, were removed to a new location an impracticable distance away. As a result of this change, everything I needed has had to be ordered and I have presented the local librarians with a number of unusual book requests, which they have been at great pains to obtain. I have also been able to use the resources of the library of the Colchester Institute and am grateful to the site librarian Ian McMeekan for his ready assistance when needed.

Another person who gave of his time to help me was Dave Mansfield, the Senior Engineer in the Colchester General Hospital. Not only did he give me a number of leads as to where I could obtain the information I was seeking but also provided me with a lot of the material I have been able to include in my manuscript. I should like to thank him for his interest.

As with the other books, I am also indebted to my friend and fellow architect Brian Roach for his continued interest in this work and for his steady flow of information, which I have found to be invaluable.

There is another group of people who must be thanked and without whom this book would never have seen the light of day. My publisher, editor, proofreader, cover designer and all the rest of the staff at Addison Wesley Longman who employed their expertise and time to translate my original manuscript into the professional book you now see. I am grateful for all their care and attention.

Above all I would like to record my deep gratitude to my wife Winifred. She has made sure that I had the time to work and, when needed, supplied the encouragement to carry on. As the preparation of a manuscript like this takes the best part of a year, such encouragement is frequently needed and, as I have been delighted to discover, not only was it generously given but Winifred also helped by discussing the text with me when I was unsure how best to explain a particular point. Without her contribution I would have found the writing of this book far more difficult.

Chapter 1

INTRODUCTION

1.1 About this book

This is the third in a trilogy of books on the subject of the Building Regulations and their attendant legislation. The first book dealt with the regulations applying to houses and flats and the second dealt with shops, offices and factories. As there are many other building types this book aims to cover the legislation applicable to all the rest. The title, *Buildings for Public Assembly and Residential Use*, attempts to describe this wide coverage and the details of the precise building types to which the book refers is given below in Section 1.3.

The basis of the book is the Building Regulations 1991 and its supporting Approved Documents. There is not a great deal of directly applicable rules set out in the Building Regulations and Documents. Instead, reference is made to the British Standards and similar guides. To ensure that the guidance in this book is as complete as it can be, the relevant Standards have been summarized as far as is practicable, to provide the information essential to the general building practitioner. It is neither possible nor really desirable to quote the whole of the British Standard but it should be recognized that, in specific circumstances, it may be necessary to obtain the full facts, in which case the original Standard should be consulted. Nor will these summaries be of great benefit to a specialist because the selected information is of a general nature and more relevant to the needs of the designer, builder and site supervisor than to the consultant.

1.2 The development of building control legislation

Historically, there have been a wide variety of forms of building control used to set standards, each with varying degrees of efficacy. In this country, control has been applied to buildings for private housing purposes for a long time but Building By-laws or Regulations relating to buildings used for the purposes covered in this book are of a more recent introduction.

The earliest, really comprehensive, building control legislation was contained in the Public Health Act of 1845. Being stimulated by public anxiety at the very low living standards in many areas of the country it, naturally, made housing its primary concern. In 1936, new legislation in the form of another Public Health Act was enacted and applied to all buildings. It was left to the local authorities to set and maintain standards and much of the by-law control was framed so that it could only apply to domestic work.

The concept of control by defining standards of performance to be achieved by the building fabric (rather than specifying how to build) was introduced in the Model By-Laws of 1952. They immediately opened the way to the introduction of new materials and technologies and effectively extended the application of control to buildings larger than houses. More particularly, they embraced the control of framed structures.

The Public Health Act of 1961 provided the basis for a national set of building regulations. These became the Building Regulations 1965 and continued the development of the control principal started in 1952. This, and a succession of amendments were consolidated into one set of Regulations in 1972 and, at the same time, the metric system of measurement replaced the imperial.

In 1974 the Health and Safety at Work, etc., Act took over the building control powers of the Public Health Acts and considerably extended the scope, purpose and coverage possible under the Building Regulations.

Three major amendments to the 1972 Regulations were made and, in 1976, a new and greatly enlarged set of Building Regulations was approved, to include the amendments and introducing a number of other significant changes. Apart from four further amendments, the first of which extended the thermal insulation requirements to embrace the type of buildings covered in this book instead of only dwellings, the next, and last, major change came about with the 1985 Regulations. These did not introduce many new requirements but totally revised the manner of presentation.

The basis set out in 1985 is still followed in the current regulations. These appear as brief functional requirements, which are the actual Regulations supported by Approved Documents. Since the latter explain, rather than state, the law, they can be written in everyday language and illustrated. Previous Building Regulations had to be written in parliamentary prose that might have been very precise in its definition but made for difficulties in understanding the requirements. The 1985 Regulations also introduced the concept of private certification but, as yet, this has only been taken up by the National House-Building Council, solely in relation to housing work.

A number of amendments to the 1985 Regulations were published in 1989 and 1991 and are contained in this book where appropriate. The 1991 amendments particularly affected dwellings by the introduction of the need to produce an energy rating of the proposal.

1.2.1 The Building Act 1984

The 1985 Building Regulations mentioned previously were made under powers conferred on the Secretary of State by the Building Act 1984.

Not only did this Act take over the enabling of Building Regulations previously contained in the Health and Safety at Work, etc, Act, it also brought together most of the legislation relating to building work that had been scattered in 46 other Acts or Instruments; the principal changes that were made affected the Public Health Acts of 1936 and 1961, the Health and Safety at Work, etc., Act 1974 and the very short-lived Housing and Building Control Act 1984.

1.2.2 Other relevant Acts

Although the Building Act 1984 brought together those parts of other enactments that impinged upon the building construction, the provision of accommodation for use as a shop, office or factory is still affected by the legislation retained in these other Acts. Principally these are:

- Offices, Shops and Railway Premises Act 1963
- Safety of Sports Grounds Act 1975
- Clean Air Act 1956
- Highways Act 1980
- Fire Precautions Act 1971
- Water Industry Act 1991
- Water Resources Act 1991
- Statutory Water Companies Act 1991
- Land Drainage Act 1991
- Control of Pollution Act 1974

As well as these Acts there are many which refer to a specific building use; where this is the case, due regard must be paid to any requirements imposed by this legislation, over and above that set out in the Building Regulations. Examples of these specific Acts are:

- The Celluloid and Cinematograph Film Act 1922 and the Cinemas Act 1985. These are administered by the local authority and, obviously, apply to cinemas.
- The Theatres Act 1968 obviously applies to theatres.
- The Licensing Act 1964 applies to any premises where alcoholic drinks are sold.
- Safety of Sports Grounds Act 1975. Under this Act, a local authority is required to issue a Certificate of Safety in respect of the entrances and exits, in which case, Sections 24 and 71 of the Building Act and any provisions under the Fire Precautions Act do not apply. (See Chapter 15.)
- Prisons Act 1952. The requirement of this Act is that the cells are certified by an inspector with regard to size, lighting, heating, ventilation and fittings

as being adequate for health and allowing a prisoner to communicate with a prison officer.

- The National Health Service and Community Act 1990 removed Crown immunity from health authorities, which means that Building Regulation Consent must be obtained from the building control authority. The National Health Service Estates publish a number of guides to hospital building under the general title of 'Firecode'; some of these are practice notes and some are Health Technical Memoranda (HTM). The Secretary of State has agreed that hospitals designed on the basis of HTM 81 or Nucleus Fire Precautions Recommendations will satisfy the requirements of Part B1 of Approved Document B to the Building Regulations, with one exception. Section 17 of Approved Document B deals with access for fire fighters. If the details in the Document are followed in the case of a hospital designed round a hospital 'street', an excessive number of stairways results. In these circumstances, the requirements are detailed in HTM 81. The principle of hospital 'street' design is further explained in Chapter 8.

The latest legislation affecting building is the Construction (Design and Management) Regulations (CONDAM). These Regulations are made under the Health and Safety at Work, etc., Act and place statutory functions on all parties concerned with construction work. They deal with planning before starting work, mainly design and construction, and continue throughout the structure's subsequent life-span. The parties concerned are the client, a planning supervisor, the designer and the contractor.

It is a duty of the client under the Regulations to appoint both the planning adviser and the principal contractor. He will probably also wish to appoint a designer, usually an architect. In appointing the contractor, the client is required to make sure that the contractor prepares a satisfactory health and safety plan prior to starting work. The planning adviser, who can be the client, the designer, the contractor or an independent person, is responsible for giving notices to the Health and Safety Executive and to advise on the appointment of the designer and the contractor if asked by the client. He must also see that the design addresses the need to reduce hazards, that relevant information concerning health and safety issues is communicated and must ensure that a health and safety file is prepared in time to form part of the tender documents so that the potential principal contractors are aware of the project's health and safety requirements. The principal contractor is responsible to see that there is adequate co-operation between all contractors to ensure compliance with statutory requirements and prohibitions and to secure the health and safety of all persons involved with the work or affected thereby. The designer is required to ensure that any design pays adequate regard to the need to avoid foreseeable risks to the health and safety of any person carrying out the construction or cleaning work at any time or who may be affected thereby.

The CONDAM Regulations reinforce the principles of the Health and

Safety at Work, etc., Act and show how they are to be applied, specifically, to construction sites.

Some local authorities, mainly county councils, have extended the provisions of certain of the other Acts to cover particular requirements perceived in their areas. These local Acts have received the approval of Parliament and may concern:

- Requirements for safety in parking areas.
- Access for fire-fighting services.
- Fire and safety precautions in large, public or high buildings.
- Separate drainage systems.
- Paving of yards and passages other than in housing.

Since these are of a local nature, they cannot be covered in this book and enquiries should be made at the local authority offices.

1.3 Building use definition

As was mentioned earlier, the aim of this book is to cover all those building types not dealt with in the other two volumes and generally described as 'buildings for public assembly and residential use'. These terms are those used in the Building Regulations as the titles of buildings whose intended use falls within these purposes. Residential Buildings are divided into two sub-groups and the Places of Assembly classification is extended to cover recreational use as well. The purpose groups are:

- Residential (Institutional)   Purpose Group 2(a)
- Residential (Other)   Purpose Group 2(b)
- Places of Assembly and Recreation   Purpose Group 5

Table D1 of Appendix D to Approved Document B describes in detail the building uses considered to fit within these general titles as follows:

Title	Premises intended to be used as
Assembly and Recreation	Broadcasting, recording and film studios open to the public
	Bingo halls, casinos and dance halls
	Entertainment, conference, exhibition and leisure centres
	Funfairs and amusement arcades
	Museums and art galleries
	Non-residential clubs
	Theatres, cinemas and concert halls
	Educational establishments, riding and dancing schools
	Public libraries
	Sports stadia, sports pavilions, gymnasia, swimming pool buildings and skating rinks

	Law courts
	Churches, other buildings of worship and crematoria
	Non-residential day centres, clinics, health centres, and surgeries
	Passenger stations and termini for air, rail, road or sea travel
	Zoos and menageries
	Public toilets.
Residential (Institutional)	Essentially, any buildings in which people live and sleep, including:
	Hospitals, nursing homes, homes for old people or children
	Schools or similar establishments used for the treatment, care or maintenance of people suffering from illness or mental or physical disability or handicap
	Place of detention.
Residential (other)	Hotel or boarding houses
	Residential college, hall of residence or hostel
	Any residential use not included above other than a private dwelling.

In addition to these purpose groups there is a Purpose Group 7, subdivided into 7(a), covering buildings used for storage and 7(b) which cover car park buildings.

1.4 Exempt buildings

Not all building work falls within the scope of the Building Act and its Regulations. Exemption arises by reason of the building's owner or intended use or by virtue of the size or temporary nature of the building.

Exempted building uses are:

- Educational buildings erected in accordance with particulars approved by the Secretary of State for Education. Recent changes in the funding arrangements for educational establishments has complicated this simple statement since not all of them have now to be so approved. Local Education Authority maintained schools, including special school projects, costing over £200,000 have to be specifically approved and are, therefore, exempt. A project of less than this value does not require specific approval and would, in consequence, be subject to the Building Regulations. However, the LEA could choose to submit it to the Secretary of State in which case it would, thereby, become exempt.
  - Self-Governing (Grant Maintained) Schools all require the Secretary of State's approval and are not subject to the Building Regulations.

- Further and Higher Education Establishments maintained by a LEA or substantially dependent for their maintenance on grants under the Education Act of 1944 are exempt from Building Regulation control. As a result of legislation in recent years, a large number of these institutions no longer fall within any of these categories and so are subject to the Regulations.
- Independent schools projects, including independent schools catering for pupils with special educational needs, are all subject to the provisions of the Building Regulations.
- Statutory Undertakers, UK Atomic Energy Authority and Civil Aviation Authority unless the building is used as an office or showroom.
- Any local authority.
- Any county council.
- Any other body that acts under an enactment for public purposes and not for its own profit and is prescribed for the purpose.

In addition, the building works shown below do not have to meet the Regulation requirements:

- Small detached buildings of less than $30\,m^2$ floor area built entirely of non-combustible materials, not less than $1\,m$ from the boundary and not containing sleeping accommodation.
- A movable building such as a tent or marquee.
- Any building not intended to remain in position for more than 28 days.
- The installation of heat-producing gas appliances provided that it is carried out by, or under the supervision of, British Gas.
- Any ancillary buildings in connection with a building site such as the site sales office, any site offices, canteen, any stores, site WCs, etc.

Note that while a contractor's on-site buildings are exempt, his permanent workshops and yard are classified as factories and subject to building control.

1.5 Definition of building work

Precisely what is a building and what constitutes building work has exercised a number of eminent legal minds. It has been held that a structure can only be a building if it has a roof but the distinction between 'structure' and 'building' is not always relevant. It would seem that a structure is not a building unless it forms part of the real estate and changes the physical nature of the land.

In more precise terms, the following have been deemed to be building works:

- The erection or extension of a building.
- Making material alterations to a building.

- The re-erection of a building when the outer walls have been pulled down or burnt down to within 10 feet of the ground.
- The re-erection of a framed building which has been so pulled down or burnt down that only the framework of the lowest storey is left.
- Roofing over an open space between buildings in excess of $30\,m^2$.
- The conversion of movable objects (mobile homes, vehicles, vessels and the like) into a permanent building,
- The provision, extension or material alteration of controlled services or fittings in a building.
- Work arising out of a change of use.

1.6 Building Regulation consent

The local authority has the duty to ensure that, when completed, building work within its area complies with the law as set down in the Building Act and the Building Regulations. Such compliance is a question of fact rather than, as in the case of planning approval, a question of opinion and judgement. The Building Regulation consent process is not whether the applicant should be allowed to build but whether the proposals satisfy the requirements and standards laid down. This makes for a significant difference in applying for consent and the consequential process employed.

1.6.1 The consent process

There are two ways to satisfy the local authority that the finished building has been constructed in accordance with the requirements of the Regulations. There is the traditional method of submitting details of the proposals before any building work is done, for the Building Control Officer to check, referred to as the full plans method or, alternatively, a developer can inform the local authority that certain works are proposed and the Building Control Officer will carry out inspections as the construction proceeds. This is referred to as the Building Notice method.

Clearly, greater risks attach to the second method since it is much easier to amend the proposals on paper before the work is done than to change the construction, if necessary, once it has been built.

For the first procedure, the necessary forms must be obtained from the local authority offices, completed in duplicate and submitted along with two copies of the drawings and a fee related to the type and extent of the work. Following their receipt, the Building Control Officer will inspect the deposited drawings for their compliance with the standards set in the Regulations. He may also call for calculations to prove the stability of the structure, the degree of thermal insulation, the capacity of the drainage system or any other aspect of the proposals requiring such proof. These will be examined by persons qualified in the appropriate skills.

The law allows the local authority five weeks in which to respond to the application. This can be extended to eight weeks with the agreement of the applicant. If, within this time, the Building Control Officer has not obtained satisfactory evidence that the Regulations will be observed he is obliged to issue a Notice of Refusal. It is not necessary, however, for the applicant to wait for the completion of this process before starting. Once 48 hours have elapsed from the time of submission of the details, work can start on site. But it must be realized that the Building Control Officer is empowered to require the removal of any work that, in his opinion, is defective or fails to comply with the requirements of the Regulations.

The second procedure can be time saving in that building work can commence before all the final details have been developed to the point necessary for submission under the first procedure. This early starting requires careful consideration to avoid the hazards of abortive work but, in certain circumstances, can be a useful advantage. Unfortunately, it is a procedure seldom available to the developers of buildings for public assembly or residential use since, with one exception, it cannot be used when the building proposals bring in a requirement for means of escape in the case of fire or affect any building that will be put to a use that has been designated under the Fire Precautions Act 1971 as requiring a fire certificate. Broadly, the buildings covered by the Fire Precautions Act Regulations are any to which the public have access or which are places of work.

The one exception allowing the Building Notice method to be used is where the proposal is for the installation of fittings such as the provision of a small washroom or toilet where such work does not affect any means of escape.

Where it is possible for this procedure to be adopted, the appropriate Building Notice form should be obtained from the local authority offices, completed and submitted. In addition to the form, copies of a plan showing the site, the proposed building or building extension, other surrounding buildings, the adjoining streets and particulars of the drainage proposals all to a scale of not less than 1:1250 must be sent in. Since this second procedure does not enjoy the benefit of a preliminary examination of the proposals by a qualified Building Control Officer, there is not the same degree of assurance as the first. It is essential, therefore, to be quite certain, not only that the proposals comply fully with the requirements of the latest Regulations but also that there are no encumbrances placed on the development of the site.

As with the full plans method, work on site can commence within 48 hours of submitting the Building Notice.

The third procedure permits the employment of the services of a person qualified to provide private certification. This was introduced in the 1985 Building Regulations but until recently it has not been taken up by anyone except NHBC Building Control Services Limited, and their services were restricted to speculative housing by registered builders. This has now changed and the DoE has now announced the approval of three companies as corporate approved inspectors for building control work, with effect from 13 January 1997. These new inspectors will be able to operate only in respect of buildings that do not contain any dwellings.

1.6.2   Fees

Fees are payable to the local authority under either of the two main procedures unless the work is solely for the purpose of providing means of access for disabled persons or to improve their safety, welfare or convenience. The scale of charges is laid down in the Regulations, with periodical amendments. Currently, the Building (Amendment of Prescribed Fees) Regulations 1992 apply; the amount payable depends on the amount of work to be carried out and is based on 70 per cent of the estimated cost.

When the full plans method is followed, the fees are charged in two parts: the first, or Plan Fee, amounting to 25 per cent of the total due, is payable when the plans are deposited and must be paid before the application will be considered. The balance of the fee, known as the Inspection Fee, becomes payable following the Building Control Inspector's first visit to see the work in progress.

Both the Plan Fee and the Inspection Fee are subject to VAT.

The fee in connection with the Building Notice method is the same in total as for the full plans method and is payable as a single sum following the Building Control Inspector's first visit to the site. The names and addresses of these companies (one in London and two in Croydon), and any others approved since this announcement, can be obtained from the local authority.

1.6.3   Inspection of the work

The Building Regulations require that either the building owner or the builder inform the local authority when certain stages in the construction have been, or are about to be, reached. This is to enable the Building Control Inspector to arrange a visit to ensure that the work is being carried out in accordance with the approved plans and the requirements of the Regulations. The stipulated stages are:

- Commencement (at least 48 hours' notice required before work starts)
- Foundation excavations
- Foundation concrete
- Damp-proof course
- Oversite hardcore
- Drains and private sewers (at least 24 hours' notice required before these are covered)
- Drains and private sewers, (notice required within 5 working days following these being covered in)

- Completion (notice required within 5 working days of finishing the building and not less than 5 working days before occupation)

Additionally, the Building Control Inspector may consider further visits are necessary, particularly in more complex buildings where he may need to check the steelwork or the floor or roof construction, for instance. Notice of these additional inspections is not mandatory under the Regulations but it assists progress of the contract and avoids having to open up completed work if additional inspections are arranged at the beginning of the job and the Inspector is informed when the appropriate stage has been reached.

Notice of inspections can be given by telephone, facsimile, in person at the local authority offices, in writing or by completing the pre-printed postcards enclosed with the notification of approval.

Failure to serve the notifications required by the Regulations could result in the local authority ordering the work to be opened up to ascertain whether it has been carried out correctly.

1.6.4 Certificates of compliance

The Building Act 1984 introduced a further change with the ruling that an 'approved person' could certify to the Building Control Officer that the proposals, or a particular part of the proposals is in compliance with the requirements of the Building Regulations. This 'approved person' is not to be confused with the 'approved inspector' required for the third method mentioned previously. The extent to which this certification can be used is limited to the stability of the structure and the ability of the fabric to conserve energy.

To act in this capacity with respect to structural stability, a person has to be approved by:

- The Institute of Civil and Structural Engineers

And for energy conservation matters the approval has to be given by:

- The Chartered Institute of Building
- The Chartered Institute of Building Services
- The Faculty of Architects and Surveyors
- The Institute of Architects and Surveyors
- The Institute of Building Control Officers
- The Institute of Civil Engineers
- The Royal Institute of British Architects
- The Royal Institute of Chartered Surveyors

In addition, the person must show that he or she has an insurance scheme that has been approved by the Secretary of State. Lists of approved persons are kept at all local authority offices.

1.6.5 Relaxation of the Regulations

The Secretary of State has the authority, under Section 8 of the Building Act 1984, to relax or entirely to dispense with the requirements of a particular Regulation in specific cases. This authority has also been delegated to the local authorities. While this might seem to be very wide sweeping power, in practice it hardly occurs at all. The reason being that almost all Regulations are now couched in terms requiring work to be 'adequate' or 'to a satisfactory level'. Any relaxation of these requirements implies permitting work that is either inadequate or to an unsatisfactory level. There is only one relaxable area and that is in Schedule 1 of Regulation B1 dealing with means of escape in case of fire. This is supported by a mandatory Approved Document and its specific requirements may be relaxed in appropriate circumstances.

In addition to these specific relaxations, the Secretary of State may also make type relaxations that are intended to relate to a particular material or technique. These type relaxations are for a limited time only, and are generally subject to certain conditions. There have been no type relaxations granted under the current legislation.

1.6.6 Type approval

To simplify the process of building control in relation to building designs that are repeated in various parts of the country, the local authorities have co-operated in forming LANTAC (Local Authorities National Type Approvals Confederation, 35 Great Smith Street, Westminster, London SW1 3BJ). LANTAC operates a national type approval scheme in England and Wales. Under the scheme a design is checked in accordance with LANTAC rules and a certificate issued stating that it complies with the requirements of the Regulations in force. The certificate is valid for three years.

Once it has received type approval a design is accepted by all the local authorities as complying with the requirements of the Building Regulations without any further examination. The approval, however, only extends to the superstructure. The siting, drainage and foundations will vary from site to site and must be individually detailed and examined.

1.7 Thermal insulation

Part L of the Building Regulations, as amended in 1994 and set out in the 1995 edition of Approved Document L, imposes standards of maximum heat loss through the fabric of a building. The principle is to limit the total heat loss via all routes, which means that a high loss through one element, say, the windows, can be offset by a low heat loss through another element, say, the roof. Three methods are given for demonstrating how heat loss through the building fabric can be limited. They are:

● The Elemental Method

- The Calculation Method
- The Energy Use Method

The first sets out standard U values for the various elements of the building which, when multiplied by the relevant areas, give a total insulation value for the whole structure.

The term 'U value' is a measure of the heat loss of a material or structure, or its capacity to transfer heat from the inside to the outside face. It is expressed in $W/m^2 K$ (watts per square metre per degree Kelvin). Watts are the measure of energy and Kelvin is used for difference between one temperature and another as opposed to Celsius which is used for the difference between zero and a given temperature. If, say, a wall possesses a U value of $0.6 W/m^2 K$ it will transfer 0.6 of a watt of heat energy through each square metre of its area for every degree of temperature difference between the outside face of the wall and the inside face. It follows, therefore, that the smaller the U value the better the insulation capacity of the wall.

The Elemental Method given in the Approved Document states that the requirement of Regulation L will be met if the thermal performance of the construction elements conforms to the following standard U values:

Roofs	$0.25 W/m^2 K$
Walls exposed to external conditions	$0.45 W/m^2 K$
Ground floors	$0.45 W/m^2 K$
Upper floors exposed to external conditions	$0.45 W/m^2 K$
Walls and floors exposed to unheated spaces	$0.60 W/m^2 K$
Windows, entrance doors and rooflights	$3.30 W/m^2 K$
Vehicle access and similar large doors	$0.70 W/m^2 K$

This set of values assumes a basic allowance for the percentage of the area of the walls and roof devoted to windows, doors and rooflights as follows:

Windows and doors in residential buildings	30% of the exposed wall area
Windows and doors in places of assembly	40% of the exposed wall area
Vehicle access doors in any building	as required
Rooflights in any building	20% of the roof area

Windows, doors and rooflights are manufactured to a range of U values which can be obtained from the producers or, alternatively, the standard values as shown in Chapter 9 can be applied. In either case, the actual value may differ from the standard value shown above, in which circumstance the percentage area allowance can be modified accordingly. Further details are given in Chapter 9.

The Calculation Method seeks to prove that the total rate of heat loss from the building would be no worse than that from a notional building of the same size and shape, designed to comply with the Elemental Method. This allows much greater flexibility in the choice of materials, construction details and window and door sizes. The Approved Document points out that although this method permits the use of walls, floors or roofs (but not all)

13

with a lower insulation value than the standard, none of these elements should be worse than $0.7\,\text{W/m}^2\,\text{K}$.

The Energy Use Method takes the principle of the Calculation Method further but, instead of proving that the rate of heat loss is no worse than that from the notional 'Elemental Method' building, it seeks to show that the calculated annual energy consumption of the building, after taking into account both heat losses and useful solar and internal heat gains, would be no worse than that of the notional building. Again, in no case should the actual U value of the walls, roofs or floors be worse than $0.7\,\text{W/m}^2\,\text{K}$.

Further reference is made to the considerations of thermal insulation in the relevant chapters dealing with each building element.

Chapter 2

MATERIALS AND WORKMANSHIP

2.1 The Building Regulation applicable

There is one complete Regulation devoted to this subject:

Reg. 7 Materials and workmanship

This Regulation requires that building work is carried out with materials that are appropriate for the circumstances in which they are used and in a workmanlike manner.

The materials referred to include all naturally occurring materials such as wood and stone, manufactured materials, components, items of equipment and any backfilling materials. The standards required are not excessive and need be no more than are necessary to secure reasonable standards of health, safety or conservation of energy. Once the building is finished the control over the use of materials imposed by this Regulation ceases.

2.2 Short-lived materials

All materials possess differing inherent durabilities and their length of useful life will also be affected by the way in which they are used. A material that may deteriorate rapidly in use if fixed in one part of the fabric could well be acceptable if located in another area. Timber, for instance, in a situation of dampness and lack of ventilation can decay or become diseased in a very short time but its employment in a position where it is kept dry or where it can be regularly treated with preservative or paint makes it fully acceptable.

The Regulations cannot set down any specific criteria from which the length of life of a material can be considered against the requirements of the Regulations. The only requirement that can be set is that the consequences of the failure of the material are not likely to have an adverse effect on the health and safety of persons in and around the building.

Clearly, a short-lived material may meet the requirements of the Regulations provided that it is readily accessible for inspection, maintenance and replacement.

2.3   Establishment of fitness of materials

While it may not be possible to set specific criteria for the length of life of a material there are recognized tests and standards that can be taken as indicative of the fitness of a material for use in a building; these are:

- The material bears an EC mark as shown in Fig. 2.1. This signifies that it complies with a harmonized standard such as a CEN or CENELEC standard mandated by the European Commission. CEN stands for Comité Européen de Normalisation, which is the European standards body that prepares harmonized standards to support the Construction Product Directive. CENELEC is the European standards body for electrical standards.

Fig. 2.1 The EC mark

- The material conforms to the relevant provisions of an appropriate British Standard. In this connection care must be taken to ensure that the correct standard is used in relation to the characteristic of the material being considered. Some standards, for instance, only refer to manufactured sizes of products and not to the durability of the material and those conforming to BS 5750: *Quality systems* can be expected to be consistent in their quality but not necessarily of adequate standard.
- The material conforms to a national standard of any Member State of the European Union or elsewhere. In this case it is up to the person selecting the material to show that in use it will be up to a relevant British Standard.
- The material is covered by an Agrément Certificate issued by the British Board of Agrément and the conditions of use are in accordance with the terms of the certificate.
- The material is covered by an equivalent technical approval of any Member of the European Union or elsewhere. In this case, as with the standards mentioned above, it is up to the person specifying the material to show that it is equivalent to a material covered by an Agrément Certificate.
- The material is certified under an independent certification scheme accredited by NACCB (the National Accreditation Council for Certification Bodies). An example of this is the Kitemark scheme operated by the British Standards Institution.
- The material can be shown by tests, calculations or other means to be

capable of performing the function for which it is intended. The National Measurement Accreditation Service run schemes for testing laboratories to ensure that tests are carried out in conformity with recognized criteria.

- The material can be shown by past experience to be capable of performing the function for which it is intended.
- Tests on samples of the material to be used show that it is of adequate standard. Regulation 7 gives the local authority the power to take and test such samples as they consider necessary to establish whether a material will comply with the requirements of the Regulations.

2.4   High alumina cement

This material was widely used in the past because of its quick-setting properties and its resistance to sulphates, but a number of building failures directly attributable to the breakdown of concrete members containing high alumina cement has led to restrictions being placed on its use. Its quick-setting properties made it an attractive material for precast concrete units because of the saving in time of use of the formwork. The first major failure of this material was in the precast concrete for beams of a school swimming pool building. Following this event a major exercise was undertaken in which all recorded uses of high alumina cement based precast units were investigated and, where it was thought necessary, safety work was carried out.

High alumina cement and any products containing it are only acceptable now in a building as heat-resisting materials. They are not permitted in any structural work, including foundations.

2.5   Asbestos

Although it is not specifically mentioned in the Building Regulations, the use of asbestos and products containing asbestos are closely controlled by the Asbestos (Licensing) Regulations 1983 because of the serious health threat it poses. It would also bring into effect the requirements of the CONDAM Regulations requiring the designer to address the potential hazards of working with the material. Asbestos dust causes asbestosis and mesotheloma – a cancer of the lung. As a result, its use in new building is now very restricted. The principal area of control over working with the material is in the removal of insulation or coatings containing asbestos or in the demolition of buildings constructed from it.

The Asbestos Regulations prohibit anyone from working with asbestos unless their employer holds a licence granted by the Health and Safety Executive, HM Factory Inspectorate, Asbestos Licensing Unit, Magdalen House, Stanley Precinct, Bootle, Merseyside LR0 3QY. A self-employed person must be similarly licensed. In addition, the Health and Safety Executive must be informed when any work with asbestos is to be carried out.

The removal of insulating board containing asbestos is not subject to the Asbestos (Licensing) Regulations but it is recommended that the same degree of care is exercised.

2.6 Softwood treatment

It is considered good practice generally to treat all structural timber with preservative but if the building is in one of a number of areas of the country, the Building Regulations require that any softwood used as roof members, including the ceiling joists and any fixed within the roof void is specially treated against attack by the House Longhorn Beetle (*Hylotrupes bajulus* L.).

The areas considered to be at risk are:

- the Borough of Bracknell Forest
- the Borough of Elmbridge
- the Borough of Guildford (other than the area of the former borough of Guildford)
- the District of Hart (other than the area of the former urban district of Fleet)
- the District of Runnymede
- the Borough of Spelthorne
- the Borough of Surrey Heath
- in the Borough of Rushmoor, the area of the former district of Farnborough
- the Borough of Waverley (other than the parishes of Godalming and Haslemere)
- in the Royal Borough of Windsor and Maidenhead, the parishes of Old Windsor, Sunningdale and Sunninghill
- the Borough of Woking

2.7 Workmanship

No matter what the quality of the materials used in building, if the workmanship is not of an equally high standard the result will be less than satisfactory. Precisely how to set and maintain quantifiable standards of workmanship is the difficult problem addressed in Section 2 of the Approved Document to Regulation 7. The way suggested is to consider a number of given aids.

The first is to check whether the method of performing the work is included in the recommendations of a British Standard Code of Practice. In this connection, BS 8000: *Workmanship on building sites* is useful in that it collects together guidance from other BSI Codes and Standards. Alternatively, the check can be extended to an equivalent technical specification of any Member State of the European Union.

The second is a similar check to see whether the subject is covered by an

Agrément Certificate issued by the British Board of Agrément or an equivalent technical approval of any other member of the European Organization for Technical Approvals or anywhere else. In the case of the latter it is up to the person proposing to use it to show that the method of workmanship will provide a similar level of satisfaction.

Thirdly, one can see whether the workmanship is the subject of a scheme that complies with the relevant recommendations of ISO 9000, EN 29000 or BS 5750: *Quality systems*. The Registration of Firms of Assessed Capability, operated by the BSI, is an example of a quality control system.

The fourth way of establishing a standard of workmanship is probably the most common, and certainly the oldest: *experience*. If it can be shown that a particular way of working has been successful in a building in use, it should be possible to assume that the completed work will be capable of performing its future function. It must always be borne in mind that this approach will only be relevant if all the circumstances are identical.

Finally, the only certain way to be sure that the workmanship is adequate is to test the work after completion. This is neither convenient nor practicable for all aspects of building but it is used for all aspects of drainage work. It is less common now for builders and architects to carry out routine tests on drains since flexible materials and more reliable jointing methods have been used, but the local authority does have the power to test installations to establish compliance with the requirements laid down for:

- sanitary pipework and drainage
- cesspools, septic tanks and settlement tanks
- rainwater drainage

Guidance on how drainage tests can be carried out are contained in the Approved Document for Part H (Drainage).

Chapter 3

WORK BELOW GROUND

3.1 The Building Regulations applicable

The Building Regulations applicable to work on and below ground are:

A1   Loading
A2   Ground movement
C1   Preparation of the site
C2   Dangerous and offensive substances
C3   Sub-soil drainage
Reg. 7 Materials and workmanship

A1 refers to the need for the building to be so constructed that the loads applied to it are safely sustained and transferred to the ground; A2 requires that the building will not suffer due to ground movement; C1 requires that the ground covered by the building must be reasonably free from any vegetable matter; C2 states that precautions must be taken against danger to health from substances present in the ground; C3 set out the requirements for sub-soil drainage if needed; and Regulation 7 is concerned with the use of suitable materials and workmanship, especially with respect to resistance to the effects of moisture and attacks by deleterious chemicals in the sub-soil.

3.2 Site preparation

To satisfy Regulation C1, the turf and any vegetable matter must be removed from the ground to a depth that will be sufficient to prevent any further growth, which means the removal of the organic top soil. If the building is one used for storage where the persons employed only take in, care for or remove the goods stored, or a building where the protection to the health of the persons working in it would not be improved by stripping the top soil, this requirement does not apply.

3.3 Site drainage

Moisture from the ground will penetrate into a building either up into the walls and floors by capillary attraction or through the structure of basements by hydrostatic pressure. There are two ways to stop this: either the moisture must be prevented from reaching the building by sub-soil drainage or a barrier must be installed to prevent its ingress.

Damp-proof courses in walls are dealt with in Chapter 4, damp-proof membranes in floors are dealt with in Chapter 5 and basement damp-proofing is covered in 3.7 below.

Approved Document C1/3 recommends that, where the water table may rise to within 0.25 m of the lowest floor of the building, sub-soil drainage should be provided unless other effective means of safeguarding the building are taken. In severe cases, the installation of sub-soil drainage in addition to moisture barriers is worth while since this will ease the pressure on the damp-proofing provision.

In the course of some building works existing sub-soil drains are cut. In this case, C1/3 requires action to be taken to prevent subsequent trouble in the building to be constructed and to maintain the existing ground conditions downstream. It is recommended that the drain is re-laid under the building in jointed pipe or re-routed round it, as shown in Fig. 3.1.

The sub-soil water is often disposed of by taking it to a ditch. Any necessary work to the ditch, such as enlarging it to take an increased flow or piping for whatever reason, may require the approval of the local water authority. Under the Land Drainage Act 1991, the Agricultural Land Tribunal may, if they think fit, make an Order giving the owner of the land the authority to enter an adjoining neighbour's property to carry out any necessary ditch work as specified in the Order.

3.4 Protection against contaminants

In this country, land available for building is, to a great extent, only in areas of existing development. This means that many sites where new building works are permitted have already been used for building purposes and some of those uses will have left a legacy of contamination requiring attention before the new building can be erected. Approved Document C defines a contaminant as any material in or on the ground to be covered by the building likely to be a danger to health and safety.

A site which is contaminated presents a potential hazard to the persons who will be working there in the course of constructing the building and, as such, the CONDAM Regulations (see page 4) will require specific measures to be included in the design and planning of the work to ensure the health and safety of all concerned.

A site surveyor will, usually, include comment on possible contamination in

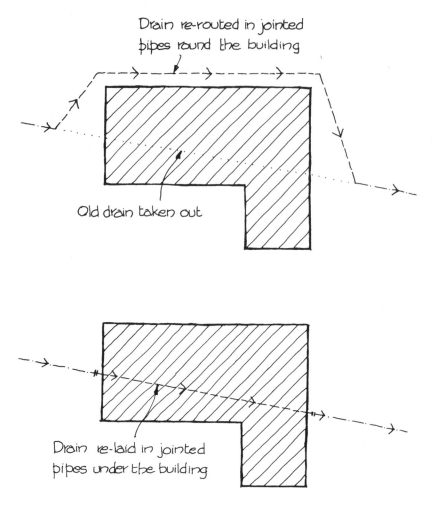

Fig. 3.1 Dealing with cut sub-soil drains

his site investigation report. This will be gained from a number of sources. One useful source is the history of the site. Both the Approved Document and the BS Draft for Development DD175: 1988 *Code of practice for the identification of contaminated land and its investigation,* referred to in the Approved Document, give lists of sites likely to contain contaminants due to earlier industrial use. The use to which areas of our towns and cities are put changes with time and it is probable that many of these former industrial sites may be considered for re-development as, say, a sports hall or hotel. For this reason a full list, made up from the two documents, is reproduced below:

● Asbestos works.

- Chemical works.
- Gas works, coal carbonization plants and ancillary by-products works.·
- Industries making or using wood preservatives.
- Landfill and other waste sites.
- Metal mines, smelters, foundries, steel works and metal-finishing works.
- Munitions production and testing sites.
- Nuclear installations.
- Oil storage and distribution sites.
- Paper and printing works.
- Railway land, especially the larger sidings and depots.
- Scrap yards.
- Sewage works, sewage farms, and sludge disposal sites.
- Tanneries.

Sites that may be radio-active are subject to special regulations and must receive especial care.

Another source of information about possible contaminants is a visual inspection. For example, vegetation may be poor, unnatural or, even, totally absent; there may be unusual colours or shapes to be seen in the surface; fumes and odours may be detectable or empty drums and containers may be lying around. All these indicate that further investigation would be prudent.

Finally, trial holes and borings can be taken and the removed sub-soil tested to establish both the presence of any offensive substances and the physical characteristics that will be required for the design of the foundations. Should any contaminants be discovered, the Environmental Health Officer must be informed. If he confirms their presence he will prescribe the action to be taken. This may consist of removal, filling or sealing.

If removal is required, the contaminant and any affected sub-soil must be taken out to a depth of 1 m below the level of the lowest floor; the requirement to fill means spreading a layer of an inert material 1 m deep over the site of the building (in this case, the type of filling and the design of the ground floor must be considered together) and to seal the site a suitable barrier must be installed between the building and the contaminant. This barrier is to be sealed at all joints, around the edges and at all service entry points. In all cases, these measures should only be undertaken with the benefit of expert advice.

3.4.1 Gaseous contaminants

In some parts of the country radon gas occurs naturally in the ground. In other areas where landfill has taken place, methane and carbon dioxide gases are to be found as a result of the action of anaerobic micro-organisms on biodegradable materials. Gases similar to landfill gas can also arise naturally.

Radon is a radio-active, colourless and odourless gas formed by the radio-active decay of uranium and radium in the soil. Exposure to high levels for long periods increases the risk of lung cancer. Methane is an asphyxiant; it will also burn and explode in air. Carbon dioxide is non-flammable but toxic. Many of the other components of landfill gas are flammable and some are toxic.

All these gases can travel through the sub-soil and migrate into the building.

Approved Document C2 gives guidance on methods of dealing with the problem with respect to housing only. For non-domestic buildings expert advice should be sought following a thorough site investigation to establish the nature, amount and pressure of any gaseous contaminants. Further information and guidance can be found in DD 175; BS 5930: *Code of practice for site investigations*; the BRE Report: *Construction of new buildings on gas contaminated land* and from the DOE publications *Notes on the development and after-use of landfill sites* and *Guidance on the assessment of and redevelopment of contamination.*

3.5 Sub-soil characteristics

To be able to determine the safest and most economic form of foundation, the characteristics of the sub-soil occurring below the level of the foundation structure must be studied, in particular, its stability and any possible causes of movement. Generally, the type and condition of the ground below a building site are such that with one of the normal forms of foundation structure the building loads will be satisfactorily carried. If, however, very difficult conditions are encountered, one of the geotechnical processes now available may be required to provide either a safe bearing or to permit the work below ground to be carried out. These processes are: injection of grout, compaction of the sub-soil, lowering of the ground water, freezing of the ground water or the use of compressed air to exclude the ground water. In all cases a specialist should be consulted.

3.5.1 Stability of the sub-soil

As the ground below the site of a building may suffer from a number of inherent causes of instability, such as geological faults, unstable strata, disused mines, etc., it is necessary to carry out a thorough site investigation before detailing any foundations for a large building. Indeed, the condition and suitability of the sub-soil may determine whether a proposed project is even feasible on a particular site. The local authority can often advise on the possibility of the existence of adverse conditions. This is a help in deciding the requirements of a site investigation.

BS 5930: *Code of practice for site investigations,* referred to in Approved Document C, states that the objects of a site investigation are to obtain reliable information to produce an economic and safe design. From the surveyor's report decisions will be made as to the most suitable system to be adopted to transfer the building loads to the ground.

3.5.2 Sub-soil movement

Movement of the sub-soil below a foundation will only occur if there is a change in its condition. The main constituent that can be changed is the water content. This can be reduced, increased or frozen and any of these events can threaten the stability of the building, particularly if the foundations are near the surface.

Reduction of the water, particularly in clay soils, will lead to shrinkage that, in turn, will cause any foundations resting upon it to move. If this movement is uniform there may be no consequential damage to the building fabric, although the external services may suffer at the point of their entry, but if it is uneven, differential movement of the building will occur with inevitable cracks or structural failure.

One way in which reduction of ground water can happen is by interference with existing land drains, which is why it is necessary to take the action described in 3.3 to maintain the level of sub-soil water downstream from the work. Similarly, the consequences of introducing new sub-soil drains should be studied.

Another, more common, way in which sub-soil water is altered is by the action of tree roots. These are capable of removing very large quantities of water from the soil and the amount increases with the growth of the tree so that, what starts as a small harmless shrub can finish up as a forest giant posing a very real threat to the stability of buildings within a wide area. The Approved Documents give no direct guidance on the siting of trees and safeguards against failure but BS 5837 suggests that, in normal circumstances, special provision should be made if the foundations are within a distance equal to the mature height of the tree. This, the Building Research Establishment recommends, should be increased to one and a half times the height if there is a row of trees. Advice should be sought in all cases as these distances need to be increased still further with certain species of tree. Additionally, large areas of paving will force trees to spread their roots further and must be taken into account.

Even when trees are removed, particularly in cohesive soils like clay, the problem persists, in this case swelling rather than shrinkage occurs due to the cessation of water extraction.

BS 8004: *Code of Practice for foundations* lists other possible causes of ground and foundation movement as:

- Seasonal changes giving rise to swelling and shrinkage of clay soils.
- Frost heave due to the freezing of sub-soil water.
- Application of artificial heat or cold that may arise due to the use of the building.

- Loss of ground due to erosion.
- Changes caused by adjacent excavations, streams, floods or the erection of adjoining structures.
- Soil creep or landslip.
- Deterioration of the ground or substructure.

3.6 Foundations

Regulation A1 is summarized at the beginning of this chapter; in full it requires that the building shall be constructed so that the combined dead, imposed and wind loads are sustained and transmitted to the ground both safely and without causing such deflection or deformation of any part of the building, or such movement of the ground, as will impair the stability of any part of another building. Regulation A2 requires that the building shall be constructed so that the ground movement caused by swelling, shrinkage or freezing of the sub-soil; or landslip or subsidence (other than subsidence arising from shrinkage), will not impair the stability of the building.

The design of the foundations, including the materials used as well as the nature of the sub-soil on which those foundations rest, must both be considered if this requirement is to be met.

3.6.1 Materials

Any material in the foundations is required by Regulation 7 to be capable of resisting attacks by deleterious substances such as sulphates. While high alumina cement is resistant to sulphates, the Regulation specifically bans its use in foundations due to other, inherent, instabilities with the material.

Approved Document A1E2 refers to BS 5328: Part 1 in which recommendations are made about the types of cement, maximum free water/cement ratios and the minimum cement contents required for differing sulphate concentrations in near-neutral ground waters of pH 6 to pH 9. Protection of the concrete is recommended where the concentration of sulphates is very high. This protection can take the form of sheet polythene or a surface coating of asphalt or chlorinated rubber.

The cements covered by the BS are ordinary Portland cement (OPC); OPC combined with ground granulated blastfurnace slag; OPC combined with pulverized fuel ash; and sulphate-resisting cement.

Concrete composed of normal Portland cement and fine and coarse aggregate can be used where the soil is not chemically aggressive, in the proportion of 50 kg of cement to not more than $0.1\,\text{m}^3$ of sand to $0.2\,\text{m}^3$ of coarse aggregate. Alternatively, it should conform to the specification given in BS 5328: Part 2 for Grade ST1 concrete.

Concrete is the usual material for strip and pad foundations and is commonly employed for pile foundations, but BS 8004 states that timber piles can also be satisfactory, particularly if the timber is used in its natural round form and is impregnated with a preservative.

3.6.2 Foundation design

As the present control of building grew out of legislation relating solely to domestic construction, the details set out in Approved Document A1E refer only to this type of work. For other forms of structure and building use, reference at the moment has to be made to BS 8004: *Code of practice for foundation design.*

The current Code, as represented by BS 8004, will be withdrawn in three or four years' time and will be replaced by Eurocode 7: *Geotechnical design.* This has been published as DDENV 1997 – 1: 1995, together with a United Kingdom National Application Document. Civil and Structural Engineers are being encouraged to design to this Code now, so that they become familiar with any changes before the Code becomes mandatory. Furthermore, as the UK document has been published, local authorities should give approval to designs based on EC7.

There are a number of important changes in principle and terminology brought about by the new Eurocode with which all building practitioners should be familiar.

Firstly, Eurocode 7 is a Limit State code. Previous design practice in this country has always been based on permissible stress principles. It is important, therefore, to ensure that the correct partial factors have been applied to the loads, materials and ground properties. Failure to do this could result in collapse.

The other main changes are:

- There are three geotechnical categories:
- Category 1 – Small and relatively simple structures like houses with a maximum wall load of $100 \, kN/m$ on conventional spread or pile foundations.
- Category 2 – Conventional structures with no abnormal or difficult ground conditions.
- Category 3 – All other structures.
- Loads are referred to as 'actions'; dead loads will be 'permanent actions' and live or imposed loads will be 'variable actions'.
- The carrying out of building or civil engineering work will be known as an 'execution'.
- Three load cases have been introduced and must each be verified:
  Case A: Hydrostatic forces.
  Case B: The strength of the structural elements.
  Case C: The strength of the ground.

One of the primary objects of a foundation design is to ensure that any foundation movement is within the tolerance of the structure placed upon it.

The different forms of construction and building size found in assembly buildings and the various types of residential building can accept equally differing degrees of foundation movement. The application of a load to the ground results in deformation and consequential settlement, which means that all buildings settle to some extent. Two forms of settlement can be identified: *immediate*, which takes place as the load is applied, and *long term*, which may continue for some time after the building is completed.

In organic soils, settlement is almost indefinite due to 'creep' in the ground and for this reason they must be avoided; silts and clays have a consolidation settlement that continues for a long time after the structure is finished, but in sands, gravels and rocks any movement will be substantially complete by the end of the construction period.

The capacity of the ground to bear loads varies from $10000\,kN/m^2$ in the case of strong igneous or gneissic rock down to less than $75\,kN/m^2$ in soft clay or silt. With this degree of difference in bearing capacity and a wide range of superstructures to be carried, it is obvious that there has to be a large range of foundation types.

There are two classes of foundations: *shallow*, which distribute the load of the building into the ground within 3 m of the surface, and *deep*, which carry the load down to a firm bearing strata. The usual forms of shallow foundation are known as spread foundations and comprise strip, deep strip, pad and raft foundations. Deep foundations take the form of pile foundations, peripheral walls and compensated foundations.

3.6.3   Strip foundations

While these are commonly used for domestic work, they are equally suitable for any buildings where the form of structure and the anticipated loading are the same as a house. This means that the construction is of loadbearing walls, the floor spans are modest and the occupancy will not impose any loading in excess of $1.5\,kN/m^2$. Generally, this loading of $1.5\,kN/m^2$ is only applicable to the bedroom and dormitory areas of clubs, colleges, institutional buildings, and residential buildings. It would not be suitable for any other use, including the bedrooms of hotels or the wards of hospitals, where a loading of at least $2\,kN/m^2$ must be allowed.

With this type of construction, the loads, which are relatively small, arrive at the foundation level uniformly distributed along a line and, therefore, a suitably sized strip of concrete placed below the wall will safely transfer the load to the ground. The wall must be located along the centreline of the concrete strip to avoid any eccentricity of loading and the width of the strip must be adjusted to take account of the type of soil. Table 3.1, based on Table 12 of Approved Document A and BS 8004 gives widths that would be suitable for the loads encountered in these 'domestic' types of building uses.

Once the width has been established, the depth of the concrete should be

Table 3.1 Minimum widths of strip foundations

Type of sub-soil	Condition of sub-soil	Field test	Minimum width (mm) for loads (kN/m) on the foundation strip of not more than:						Undrained shear strength BS 8004 (kN/m²)
			20	30	40	50	60	70	
Rock	Not inferior to sandstone, limestone or firm chalk	Requires at least a pneumatic pick for excavation	Equal to the thickness of the wall						Strong — 4000
Gravel, Sand	Compact	Requires a pick to excavate; 50 mm square peg hard to drive beyond 150 mm	250	300	400	500	600	650	Dense — over 600; Medium dense — 200–600
Clay, Sandy clay	Stiff	Cannot be moulded in the fingers; removal requires a pick or pneumatic spade	250	300	400	500	600	650	Stiff — 100–150; Firm to stiff — 75–100
Clay, Sandy clay	Firm	Can be moulded by substantial finger pressure; can be dug with a spade	300	350	450	600	750	850	Firm — 50–75; Soft to firm — 40–50
Sand, Silty sand, Clayey sand	Loose	Can be excavated with a spade; 50 mm square peg easily driven in	400	600	Not applicable if the load per metre exceeds 30 kN. Strip widths may be calculated but expert advice is needed				Less than 200
Silt, Clay, Sandy clay, Silty clay	Soft	Fairly easily moulded in the fingers and readily excavated	450	650					20–40
Silt, Clay, Sandy clay, Silty clay	Very soft	Natural winter sample exudes between the fingers when squeezed in the fist	600	850					Less than 20

(Source: Based on Table 12 of Approved Document A1 and BS 8004)

set so that it is not less than the distance between the wall face and the edge of the concrete strip. This is to allow the imposed stress to spread through the concrete at an angle of 45°, as shown in Fig. 3.2.

The level at which the strip is placed in the ground is determined by the nature of the soil and the need to avoid movement. BS 8004 states that in clay soils the strip foundations of a traditional brick building should be founded at a minimum of 900 mm below the finished ground level, and more often than not this is rounded up to 1 m. A further increase on this depth is necessary if the site is likely to be affected by tree roots.

The widths given in Table 3.1 relate to buildings with a loading not exceeding 1.5 kN/m². Other buildings, built in the traditional manner with loadbearing brick walls, can still be founded on a concrete strip but the width must be calculated from the formula:

$$\text{Minimum width} = \frac{\text{Total wall load per metre}}{\text{Allowable bearing pressure}}$$

The same rules regarding the depth of the strip and its position in the ground apply.

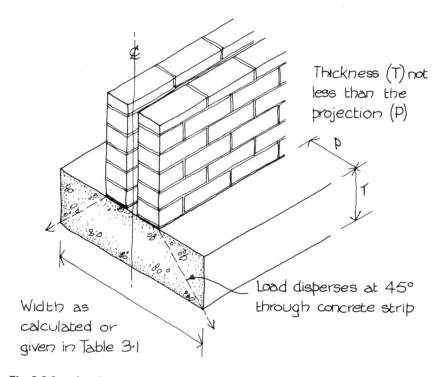

Fig. 3.2 Strip foundation

3.6.4 Deep strip foundations

Sometimes called trench fill foundations, the principle is the same as strip foundations with the difference that instead of a strip of concrete supporting a foundation wall, the whole trench is filled with concrete to a level just below ground. The advantage is one of economy. In many cases the excavated width of a strip foundation has to be greater than the minimum required for loadbearing purposes to allow space for the bricklayer to work when constructing the foundation wall. If the trench is filled with concrete, this need is avoided and the minimum loadbearing width can be used. This is advantageous in soils with a good bearing capacity.

3.6.5 Pad foundations

In many buildings used for assembly purposes in particular, the required size of the internal spaces requires that a framed structure must be adopted. This form of building produces concentrated loads where the columns occur, making a continuous, strip foundation inappropriate. The solution is to provide pad foundations where the magnitude of the loads and the bearing capacity of the ground will allow.

These are square slabs of concrete, which may or may not be reinforced, placed centrally under each loadbearing point and sized to ensure that the load is distributed safely into the ground. The area required is found by simply dividing the load by the bearing capacity of the ground. For example, if the load on a column is 200 kN and the bearing capacity of the ground is 50 kN/m^2, the area of the pad foundation that would safely sustain this load would be 4 m^2, i.e. 2 m $\times$ 2 m square.

The required thickness is found from the same rules that apply to a strip foundation – that is, the distribution of the load is assumed to be at 45° from the edges of the column base through the pad and must be contained within the concrete. It follows that the larger the slab the thicker it has to be, and a point is reached when it is more economical to use a calculated reinforced concrete pad.

The depth of the underside of the pad below ground level is also controlled by the same considerations as strip foundations with regard to the nature of the sub-soil and the possibility of movement due to frost and other causes.

3.6.6 Raft foundations

If the size or spacing of foundation pads leave very little space between them it is probably more practicable to join them up into one slab to form a reinforced concrete raft foundation. The shape of the raft should be as near to square as is possible to try to ensure a uniformity of stress throughout. Its position in the ground can be relatively near the surface but, if it is, precautions should be taken to minimize the effect of frost penetrating under the edges, causing large movements. This type of foundation requires the advice of a specialist.

3.6.7   Pile foundations

BS 8004 defines three types of pile foundation, large displacement, small displacement and replacement (see Fig. 3.3).

Large displacement piles are driven piles of substantial section, they may be of timber, solid concrete or a concrete tube closed at the bottom end with a shoe. In being driven into the ground, their volume must be accommodated by both vertical and lateral displacement of the soil, particularly if there is a group of piles when the displacement may be considerable. This heave may affect piles already driven and it can disturb adjoining buildings. The noise and vibration generated by the operation of pile driving may also be unacceptable in the neighbourhood. The advantage of the system is that the force required to drive the pile gives an accurate indication of its loadbearing capacity.

Small displacement piles are made up from rolled steel sections, H piles, open-ended tubes or hollow sections. Since there is not the same cross-sectional area as solid sections described above, the displacement caused by driving these piles is significantly less. The noise and vibration problems still exist and, in certain circumstances, their bearing capacity will be less than a solid section pile.

Replacement piles are the bored type where the ground is removed by an auger and the resulting shaft filled with reinforcement and concrete.

The great advantage of bored piles is that no vibration is caused and the level of noise is much less than with driven piles. Problems can be encountered in loose ground where the side of the bore hole tends to fall in before the pile can be completed and sub-soil water in quantity can disrupt operations. In both cases, the use of a hollow casing that follows the auger down into the ground is needed to prevent any inflow. As is prudent with all subterranean work, the concrete must be of a sulphate-resistant type.

Whatever form of pile is used, the working load from each column in the building is transferred via a capping plate fitted to the top of each pile or, where the loads are high or the bearing capacity low, to a group of piles.

3.6.8   Ground beams

Pad foundations and pile systems are the common methods of supporting the point loads encountered in framed buildings, but the columns creating these point loads are not always the only structural elements requiring support, the frame has to be clad and the enclosed space, in some schemes, subdivided into rooms. Where appropriate, the frame cladding can be fixed to rails running from column to column but in a lot of assembly and residential

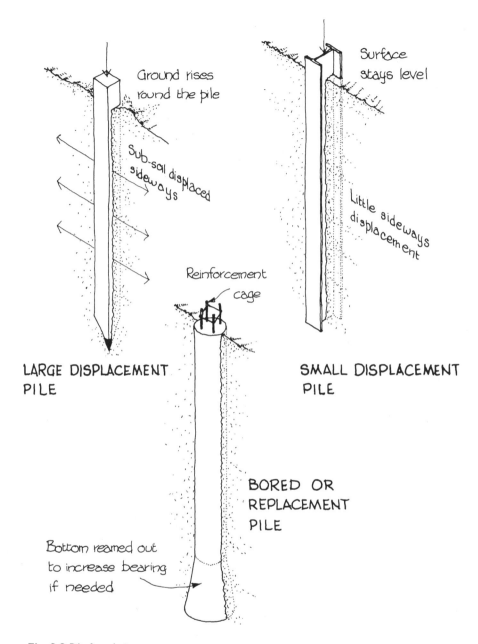

Fig. 3.3 Pile foundations

buildings the enclosure is of a more substantial nature and the provision of a support structure at ground level becomes necessary.

The usual method of providing support for enclosing and subdividing walls is by the construction of ground beams (see Fig. 3.4). These are reinforced

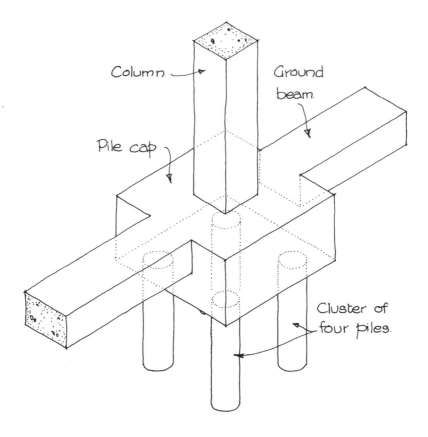

Column → Ground beam.

Pile cap

Cluster of four piles.

Fig. 3.4 Pile cap and ground beams

concrete beams cast in trenches in the ground and spanning from one column foundation to another. By this means, all the loads are placed at preselected points via a linked foundation structure and the risk of differential movement between the frame and the walling is considerably reduced.

3.6.9 Peripheral walls

In many buildings economic or land usage factors or planning restrictions may dictate the use of below ground accommodation. Swimming pools in sports complexes are also, frequently, set down into the ground but the design and construction of these is a specialized operation, not covered in the Building Regulations.

The first difficulty to be solved when contemplating basement construction is how to maintain a large hole in the ground while the structure is being built. One approach, mentioned in BS 8004, is to form the basement-enclosing structure before excavating the area, thereby providing the necessary support to the sides of the excavation to prevent them falling in. Two principal methods exist: diaphragm walling and contiguous piling (see Fig. 3.5). Both of these will provide a foundation to support the superstructure of the building as well as retain the adjacent ground. Indeed, in many designs,

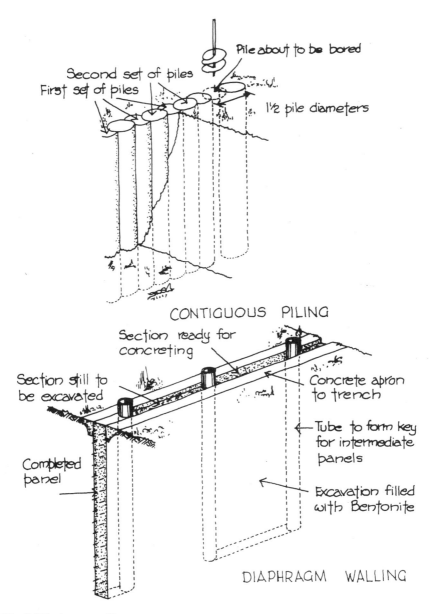

Fig. 3.5 Perimeter walling

the load of the superstructure acting vertically is relied upon to counter-balance the horizontal thrust of the retained ground.

Diaphragm walling is constructed by excavating a very narrow trench, equal in width to the required thickness of the diaphragm wall and as deep as the finished basement, lowering a cage of reinforcement into it and then filling it by pumping in concrete. The technique is made possible by the use of a slurry, usually a thixotropic clay like Bentonite, which is run into the trench as it is being excavated and then stiffens to prevent the sides from collapsing. The reinforcement can be placed into the slurry-filled trench without difficulty and as the concrete is pumped in it displaces the slurry which is then collected for re-use elsewhere. Once the concrete has cured to its working strength the ground within the diaphragm walling can be dug out and the construction completed as outlined in Section 3.7. There are obvious limits on the depth to which this construction can be carried, partly determined by the reach of excavating machinery but also depending on the shape and size of the building, the nature of the ground and the horizontal loads the wall will be required to support as the main excavation proceeds.

Contiguous piling can be taken to a deeper level more readily than diaphragm walling as it is composed of bored piles. Unlike bored piles for isolated foundation bases, contiguous piles are set so close together that they link up to form a continuous structure. The process is to take out a row of bore holes spaced at intervals of slightly less than the diameter of the piling and in them form reinforced piles. Before the concrete of these piles has set fully, another set of bore holes is made between each pair of piles. In the process, part of the previously cast piles is bored away. When the second set of piles is formed they interlock with the first to form the required peripheral wall. As with diaphragm walling, contiguous piling is usually designed to act both as a foundation to the superstructure and a retaining wall to the ground.

Once the piling is installed and the concrete cured the main excavation and completion of basement construction can proceed.

3.6.10   Compensated foundations

The excavation of the soil when a basement is formed removes a large load that, formerly, bore down on the sub-soil below the bottom of the excavation. The principal of compensated foundations is that, if the right amount of sub-soil is removed, the mass of the excavated material will equal the mass of the building and, if uniformly distributed over the area of the excavation, will not impose any greater stress than that which existed before work commenced. By this means the risk of settlement is almost entirely abolished (there is a certain amount of movement because when the mass of sub-soil is taken away the remaining ground expands upwards and will return to its original volume when the building loads come onto it).

Tall buildings are suitable for this type of treatment, particularly if the use of the basement void allows it to be subdivided into a number of small cells to enable the building loads to be distributed over the area of the base of the excavation.

The construction may use either diaphragm walls or contiguous piling or it may be formed by the use of a caisson. This technique, as mentioned in BS 8004, is more commonly employed in engineering work, especially for bridge piers but, if the ground conditions are very wet or loose, it may be an economic answer for building work.

The caisson is a reinforced concrete shell, usually circular in plan but for building purposes may be rectangular, either precast or cast in-situ and positioned on the ground where the excavation is to be made. With the shell in place the soil within and below it is then dug out and the caisson allowed to sink into the ground under its own weight. Further sections can be added to the top and the process is continued until the desired level is reached.

3.7 Basement construction

Once having obtained a large clear hole in the ground by one of the means outlined in the previous sections, or by the use of temporary sheet steel piling, the basement must be completed to the appropriate standard. If a diaphragm walling or contiguous piling system is used, no further side supporting structure is needed. With temporary piling a permanent structure of either brick or reinforced concrete must be constructed within the excavated area. In both cases a floor will have to be laid which is capable of resisting the upward forces likely to be experienced.

In addition to the structural considerations, damp-proofing of the basement must also receive attention in most buildings. Approved Document C4 gives technical details of damp proofing of floors laid on the ground, but for floors and structures below ground it refers to Clause 11 of CP 102: *Protection of buildings against water from the ground* and to BS 8102: *Code of practice for protection of structures against water from the ground*. The latter deals with the problems of floors and walls subject to water under pressure.

Three types of system are described in BS 8102 as being suitable to withstand hydrostatic pressure:

- Type A   Tanked
- Type B   Structurally integral protection
- Type C   Drained protection

Type A consists of a concrete or masonry construction built within an excavated area. This, of itself, offers no resistance to the penetration of either water or water vapour, so a continuous barrier system is applied that provides the necessary protection (see Fig. 3.6). This barrier, or tanking as it is usually called, can be carried out with:

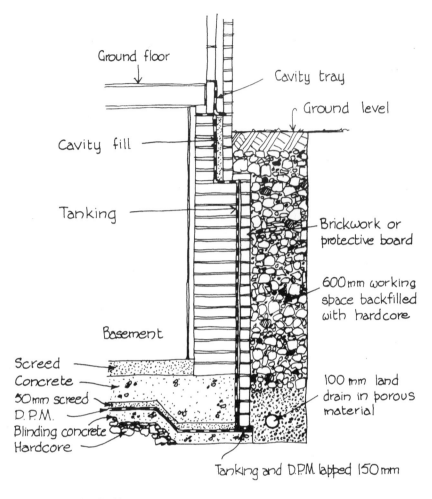

Fig. 3.6 Type 'A' tanked basement protection

- mastic asphalt
- preformed bitumen sheeting
- proprietary sheeting
- cement-based waterproof renders (but this will allow water vapour through)
- polymer/cement-based waterproof coating (two or more coats)
- resins of polyurethane or similar materials
- high performance bonded elastomeric tanking membranes.

The tanking is best applied externally because then the force of the ground water presses the tanking against the structure, but where space on the site does not permit access to the external face it can be applied internally. If the tanking is internal a loading coat must be provided to prevent the hydrostatic pressure from rupturing the tanking by pushing it away from the structure. Special measures must be taken wherever there are movement joints or where pipes penetrate the tanking.

For Type B, BS 8102 refers to two other British Standards, BS 8110 for reinforced or prestressed concrete constructions that minimize water penetration and BS 8007 for reinforced or prestressed work that prevents water penetration. In neither case is the structure capable of preventing some penetration of water vapour.

Type C is the most effective and trouble free and consists firstly of a concrete structure that minimizes the penetration of water. This is backed up by an inner construction that creates a cavity at least 75 mm wide which collects any moisture passing through the concrete and channels it away to a sump (see Fig. 3.7). Vapour transmission is prevented by ventilating the cavity in the walls and laying a damp-proof membrane below the floor.

In all cases the amount of water to be dealt with can be reduced by surface paving laid to falls for a distance of 3 m out from the building to prevent the rainwater soaking in and by the provision of sub-surface drainage.

The choice of which of the three systems to adopt should be based on the use to which the basement is to be put and the standard of dryness necessary, as shown in Table 3.2.

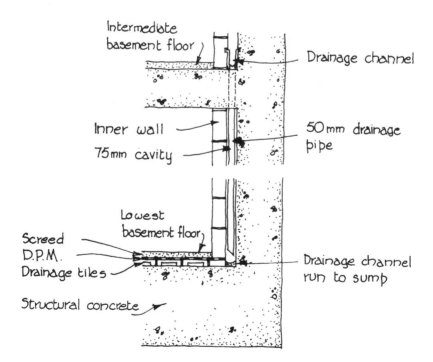

Fig. 3.7 Type 'C' drained cavity basement protection

Table 3.2 Standards of construction of basements

Use of basement space	Standard of dryness required	Suitable type of construction
Car parking Plant Room (not electrical) Workshops (not requiring a dry environment)	Some seepage and damp patches acceptable	Type B – with reinforced concrete to BS 8110 (minimizes water penetration)
Workshops and Plant Rooms (requiring a drier environment) Retail storage areas	No water penetration tolerable Moisture vapour acceptable	Type A – tanked protection or Type B – with reinforced concrete to BS 8007 (prevents water penetration)
Ventilated working areas Offices Restaurants Leisure centres	Reasonably dry environment	Type A – tanked protection or Type B – with reinforced concrete to BS 8007 (prevents water penetration) or Type C – drained protection plus d.p.m. in floor
Archives and stores (requiring controlled environment)	Totally dry	Type A – tanked protection or Type B – with r.c. to BS 8007 plus vapour barrier or Type C – drained protection plus ventilation to cavity, vapour barrier to inner skin and d.p.m. in floor

(*Source:* Based on Table 1 of BS 8102)

3.7.1   Means of escape from basements

Basements present a higher fire risk than upper storeys as there is no external access; furthermore, a fire in this region would endanger the whole of the building. For these reasons, special measures are required for means of escape and to protect the upper storeys.

The minimum stair width should be determined on the basis of total evacuation, this means that the stairs should be wide enough to allow all floors below ground to be evacuated simultaneously (for minimum width calculations, see Chapter 15). In addition, if the whole building is served by one escape stair only, the basement must be served by a stairway that is separate from the main escape stair. Where more than one escape stair is provided to the upper storeys, only one needs to terminate at ground floor level, the others can continue down into the basement if desired but must be separated from each basement storey by a ventilated protected lobby or corridor.

3.7.2   Smoke venting of basements

A fire in a basement tends to fill the stairway with smoke, making access difficult for the fire-fighting personnel. Smoke also reduces vision, making search, rescue and fire fighting more difficult. Section 18 of Approved Document B5 gives guidance on the provision of smoke vents to assist the fire service personnel and this is dealt with in detail in Chapter 15.

Chapter 4

EXTERNAL WALLS

4.1	The Building Regulations applicable

The Building Regulations relating to the construction of external walls are contained in the following; in addition, reference is made in the Regulations to BS 6399: *Loading for buildings*; BS 5628: *Code of practice for use of masonry*; BS 8110: *Structural use of concrete*; BS 5950: *Structural use of steelwork in buildings*; BS 449: *Specification for the use of structural steel in buildings*; CP3: *Code of basic data for the design of buildings*; and CP 118: *The structural use of aluminium.*

A1   Loading
A3   Disproportionate collapse
B2   Internal fire spread
B4   External fire spread
C4   Resistance to weather and ground moisture
D1   Cavity insulation
L1   Conservation of fuel and power
Reg. 7 Materials and workmanship

A1 requires that the building will safely sustain the dead loads, the imposed loads and the wind loads to which it is subject; A3 relates to buildings over five storeys high and states that in the event of an accident any resultant collapse shall not be disproportionate to the cause; B2 is concerned that the rate of spread of fire within a building shall be inhibited by the resistance to spread of flame by the internal linings and, if ignited, the rate of heat release is reasonable; B4 states that the external walls of a building must resist the spread of fire from one building to another; C4 deals with provisions to keep the interior of the building dry and the Approved Document includes a section specifically related to the cladding of framed buildings; D1 is about the prevention of toxic fumes from cavity insulation finding their way into a building; L1 requires that reasonable provision is made to conserve fuel and power in buildings with a floor area over $30 \, m^2$ and Regulation 7 deals with materials and workmanship to be employed.

In addition, a set of Building Regulation Approved Documents includes a copy of the BRE Report *Thermal insulation: avoiding the risks*. This is not an Approved Document but its recommendations are given where appropriate.

4.2   External walls in small buildings

Generally the guidance given in Approved Document A1 is intended to apply to domestic buildings with the exception of small, single-storey buildings for non-residential use. By 'small' the Approved Document means a building not exceeding 3 m to its highest point and not exceeding 9 m wide on normal or slightly sloping sites. Where the site is steeply sloping, such as in a cliff or escarpment location and there is no natural protection for the building, the guidance applies only if the basic wind speed is less than 44 m/s (see Fig. 4.1).

Gust speeds likely to be exceeded only once in 50 years at 10 m above open level country

Fig. 4.1 Map of basic wind speeds

4.2.1 Wall thickness

Two rules are given for solid walls, both producing almost the same result. Firstly, the wall must not be less than 1/16 of the storey height, and with a maximum of 3 m this gives a thickness of 187.5 mm. Secondly, Table 5 of Approved Document A1/2 shows that the minimum thickness for a wall not exceeding 3.5 m high or 12 m long (a 'small' building) is 190 mm for the whole of its height.

If the walls are of cavity construction, each leaf must be at least 90 mm thick and the cavity not less than 50 mm wide. Cavity ties, generally, must be placed at 900 mm horizontally and 450 mm vertically. If the cavity is 76 to 100 mm wide the horizontal spacing must be decreased to 750 mm.

Notwithstanding these rules, it is permissible to construct the external walls of small, single-storey non-residential buildings with a single leaf of brickwork not less than 90 mm thick. Where these walls do not enclose a floor area exceeding 36 m² and are not subject to any load other than from the roof or the wind, they should be supported by 190 × 190 mm piers or buttress walls at a spacing of 3 m or less, as shown on Fig. 4.2.

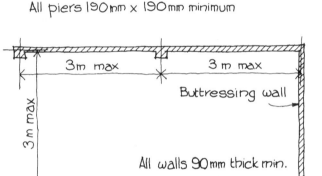

Fig. 4.2 Controlling dimensions of a 'small building'

4.2.2 Materials

In giving this guidance for small buildings, it is assumed that the compressive strength of any bricks used for the outer skin of a cavity wall should be not less than 5 N/mm² and for blocks in the inner skin, not less than 2.8 N/mm². Solid walls should be built of bricks or blocks with a compressive strength of not less than 2.8 N/mm². This figure of 2.9 N/mm² is the minimum crushing strength for lightweight blocks conforming to BS 6073. Bricks vary widely in their strength but even the weakest common bricks will withstand 7 N/mm² and facings may start as high as 20 N/mm².

The mortar to be used should be 1:1:6 Portland cement, lime and sand or to the proportions given in BS 5628 for mortar designation (iii).

4.2.3 Loadings

The imposed load due to the roof is assumed in Approved Document A1/2 to be 1.5 kN/mm^2 where the span is less than 6 m and 1.0 kN/mm^2 for spans over 6 m; for ceilings it is taken as 0.25 kN/mm^2 together with a 0.9 kN concentrated load. These loadings must be uniformly distributed along the wall.

Wind loads are related to the basic wind speed and the height of the building. Since the height of these small buildings is restricted to 3 m, the guidance on wall thickness can be applied in all localities with any wind speed (as shown on Fig. 4.1) with one exception. Where the site slopes steeply and the building has no natural protection other guidance must be sought if the basic wind speed is expected to exceed 44 m/s.

4.3   Structure and loading

As the guidance given in Approved Document A1/2 applies only to residential buildings or the small non-residential buildings described above, any further guidance on meeting the requirements of the Regulation must be found in the relevant British Standards. The notes in the following sections are all based on the Standards as shown.

Regulation A3 relates particularly to framed structures and requires that an accidental damage will not give rise to a collapse that is disproportionate to the cause. This requirement arose originally as a consequence of the collapse of a block of flats at Ronan Point where the complete corner of the multistorey building came down following the failure of one panel joint.

The requirement can be met, firstly, by adopting measures that will reduce any hazards in the building that may give rise to the kind of accident that could seriously damage the structure. Many of these measures will be dictated by specific rules or regulations concerning the precise use of the building and, for instance, the precautions to be taken over the storage of materials.

The second means of meeting the requirements is to reduce the sensitivity of the structure to disproportionate collapse. Section 5 of Approved Document A3/4 shows that this would be satisfied by the provision of effective and adequate horizontal and vertical ties. It is also recognized that it may not always be possible or feasible to provide such tying and alternatives are offered.

If it is not possible to provide vertical ties in the structure, then each loadbearing member must be considered to have been removed in turn in each storey and a check applied that the remaining structure will bridge over the missing member, albeit in a substantially deformed state. Any areas of the structure that would, inevitably, collapse as a result of such removal (cantilevers or simply supported floor panels, for instance) must be limited to 15 per cent of the area of the storey and the immediately adjacent storey, or $70\,\text{m}^2$, whichever is the less. It should be noted that the area referred to is not necessarily the area taken in the structural calculations as being supported by the member under examination but the area that would collapse (see Fig. 4.3).

A further option is that if it is not possible or feasible to provide either vertical or horizontal ties then each loadbearing member must be considered

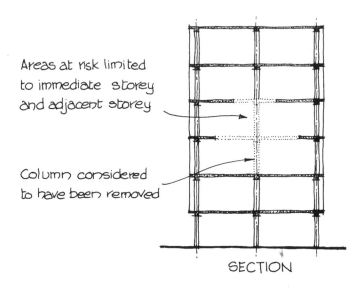

Areas at risk limited to immediate storey and adjacent storey

Column considered to have been removed

SECTION

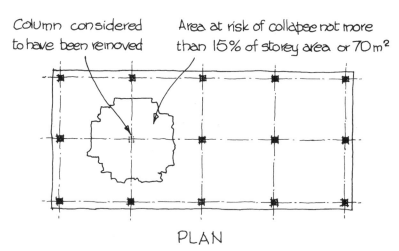

Column considered to have been removed

Area at risk of collapse not more than 15% of storey area or $70\,\text{m}^2$

PLAN

Fig. 4.3 Limitations of theoretical collapse

as having been removed in turn as described above and the consequential collapse limited to 15 per cent of the storey or $70\,m^2$.

In all the options given, if it proves impossible to achieve a satisfactory limit on the extent of the collapse, then the member concerned should be a 'protected member' or 'key element', designed in accordance with the recommendations given in:

- BS 5628: Part 1, in the case of masonry buildings
- BS 5950: Part 1, in the case of steel-framed structures
- BS 8110: Part 1, in the case of reinforced concrete structures

4.4 Structural work of timber

Approved Document A1/2 gives no guidance on the construction of external walls in non-residential buildings other than to refer to BS 5268: *Structural use of timber.* Part 2: *Code of practice for permissible stress design, materials and workmanship.* The following notes are all taken from this reference. The Standard covers the structural use of timber, plywood, glued laminated timber, tempered hardboard and wood chipboard including the design of nailed, screwed, bolted and glued joints. It should be noted, however, that a new European Discussion Document has been published, DD ENV 1995–1: 1994 Eurocode 5: *Design of timber structures,* accompanied by the UK National Application Document. Although this is not yet the standard the Regulations require structural engineers to follow, they are encouraged to design to this code so as to become familiar with the different principles involved. It is probable that the current BS will be withdrawn by the end of the millennium and Eurocode 5 will take its place.

The main differences between the BS 5268 and Eurocode 5 are:

- Designs are to be based on a Limit State code rather than, as at present, permissible stress.
- The naming of axes is to change from X–X for the horizontal axis and Y–Y for the vertical axis as now, to Y–Y for the horizontal axis and Z–Z for the vertical (see Fig. 4.4),
- 'Loads' are to be referred to as 'actions'.
- 'Dead loads' will become 'permanent actions'.
- 'Imposed loads' will be called 'variable actions'.
- The reduction in total imposed floor load described in BS 6399: Part 1, clause 5 is not permitted under Eurocode 5.
- Wind load should be taken as 90 per cent of the value obtained from CP3: Chapter V: Part 2.
- Two deflections are considered: instantaneous and final deflection, including creep.
- New timber grades will be phased in, C16, which is approximately the same as SC3 and C24 which approximates to SC4 (strength grades are explained below).

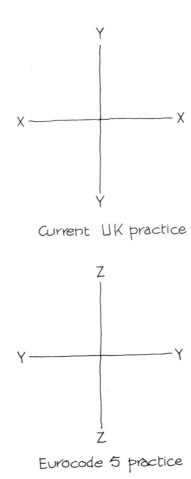

Current UK practice

Eurocode 5 practice

Fig. 4.4 Change of axis nomenclature

- Instead of restraints on the breadth to depth ratio, there is to be a new 'instability factor' which relates to lateral torsional buckling.
- Finger joints are to carry a third party quality assurance by a body approved by the National Accreditation Council of Certification Bodies.

Most of these changes will only affect structural engineers in the calculation of a structural frame but the different axis nomenclature, the renaming of loads and the requirements for finger joints are all changes about which everybody in the industry should be aware.

As a principle, any timber structure must sustain the dead, imposed and wind loads with due stability and stiffness. To achieve this, BS 5268 states consideration must be given to:

- The geometry of the structure.

- Required interaction and connections between the timber loadbearing elements and between those elements and other parts of the structure.
- Bracing and diaphragm effects in planes parallel to lateral forces acting on the structure.

The design must give a reasonable probability that the structure will not collapse catastrophically as a result of misuse or accident, nor be damaged to an extent disproportionate to the cause. This last point in the Standard on disproportionate collapse corresponds exactly to the requirements of Regulation A3. While there are no specific recommendations in the Standard for buildings up to four storeys, the principles should always be considered.

The three main points to examine are:

- The effect of an exceptional load due to drifting of snow on the roof (this should be checked regardless of height). Under Eurocode 5, snow loads arising from drifting would be treated as an 'accidental action' and of short duration.
- The effect of the hazards of a particular occupancy to ensure that, in the event of an accident, the structure will continue to perform its function even if damaged.
- The consequence of vehicles running into vital loadbearing members on the ground floor. These should be protected by bollards, walls, earth banks or the like.

While the structure should be designed so that the loads normally induced do not exceed the permissible values given in the Standard, when considering the effects of misuse or accident, or when assessing the residual stability of a damaged structure, these permissible stresses and loads on fasteners can be multiplied by two.

4.4.1 The permissible stresses used in the design of timber frames

The design of timber frames is fundamentally simple but with a lot of refinement by the use of modification factors. There are 84 modification factors, all with the letter K plus a suffix from 1 to 84 and all relating to various physical or geometric properties of the material (additional factors will be introduced by the new Eurocode 5 as mentioned above). The calculation of the stresses occurring in a timber frame is a specialized task, better left to a qualified engineer and cannot be included in this book. If the actual permissible stress values are required, reference should be made to the BS.

Initially the design must start with the permissible stress the timber can accept. This is derived from the actual basic stress modified by the appropriate factor. Each species of structural timber is now stress graded in accordance with BS 4978 and classified as GS (General Selected), SS (Special Selected), MGS (Machine General Selected), MSS (Machine Special Selected), M50 or M75. These stress grades are translated into strength classes that

Table 4.1 Combination of species of softwood and strength grade (to BS 4978) which satisfy strength classes SC1 to SC5

Standard name	Strength class				
	SC1	SC2	SC3	SC4	SC5
British grown timbers					
Douglas fir		GS	M50 or SS		M75
Larch			GS	SS	
Scots pine			GS or M50	SS	M75
Corsican pine		GS	M50	SS	M75
European spruce	GS	M50 or SS	M75		
Sitka spruce	GS	M50 or SS	M75		
Imported timbers					
Redwood/Whitewood			GS or M50	SS	M75
Douglas fir – larch from Canada or USA			GS	SS	
Hem-fir from Canada			GS or M50	SS	M75
Hem-fir from USA			GS	SS	
Spruce-pine fir from Canada			GS or M50	SS or M75	
Sitka spruce from Canada		GS	SS		
Western whitewoods from USA	GS		SS		

(*Source:* Based on BS 5268: Part 2)

are numbered SC1 to SC5 (see Table 4.1). Basic grade stresses are assigned to each strength class as shown in Table 4.2.

Glued laminated members have their own sets of permissible stresses. They should be manufactured in accordance with BS 4169 using stress-graded timber and can have the laminates arranged horizontally or vertically (see Fig. 4.5).

Curved laminated members should be designed so that the ratio between the radius of curvature (r) and the thickness of each lamination (t) is greater than 125 for softwood and 100 for hardwood.

4.4.2 Joints

The permissible loadings for connections in timber frames are tabulated with reference to the stress class of the timbers joined. Any fasteners to be used in

Table 4.2 Grade stresses and moduli of elasticity for strength classes

Strength class	Bending parallel to grain	Tension parallel to grain	Compression		Shear parallel to grain	Modulus of elasticity		Approx. density
			Parallel to grain	Perpendicular to grain		Mean	Minimum	
	(N/mm^2)	(N/mm^2)	(N/mm^2)	(N/mm^2)	(N/mm^2)	(N/mm^2)	(N/mm^2)	(kg/m^3)
Softwoods								
SC1	2.8	2.2	3.5	1.2	0.46	6800	4500	540
SC2	4.1	2.5	5.3	1.6	0.66	8000	5000	540
SC3	5.3	3.2	6.8	1.7	0.67	8800	5800	540
SC4	7.5	4.5	7.9	1.9	0.71	9900	6600	590
Softwoods or hardwoods								
SC5	10.0	6.0	8.7	2.4	1.00	10700	7100	590/760
Hardwoods								
SC6	12.5	7.5	12.5	2.8	1.50	14100	11800	840
SC7	15.0	9.0	14.5	3.3	1.75	16200	13600	960
SC8	17.5	10.5	16.5	3.9	2.00	18700	15600	1080
SC9	20.5	12.3	19.5	4.6	2.25	21600	1800	1200

(*Source*: Based on Table 9 of BS 5268: Part 2)

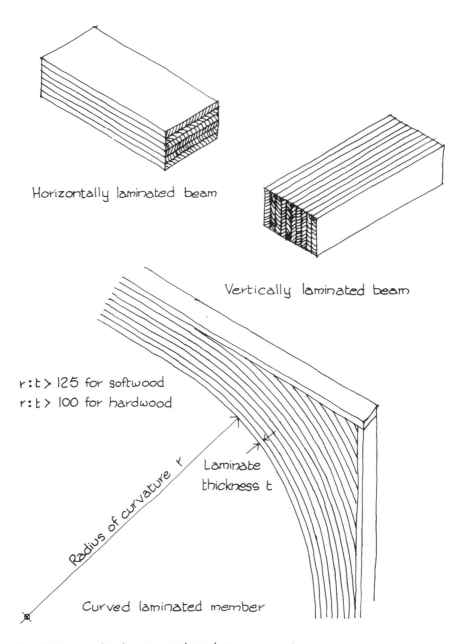

Horizontally laminated beam

Vertically laminated beam

r:t > 125 for softwood
r:t > 100 for hardwood

Radius of curvature r

Laminate thickness t

Curved laminated member

Fig. 4.5 Laminated timber structural members

wet timber, or in conditions where they may be expected to get wet, must be either non-corrodible or treated by an anti-corrosive process.

Nails should be manufactured in accordance with BS 1202: Part 1.

Any nailed joint should contain at least two nails. While the stress values allowable in nailed joints are given for round wire nails driven into the timber at right angles, the Standard does mention that skew nailing increases the withdrawal resistance of the nail but warns against skew nailing where reversal of the loading could occur. The skewing of the nail should be arranged so that, under load the joint will tighten and opposed skewing is preferable to parallel skewing, see Fig. 4.6.

If nails are placed too near the end or the edge of the timber or are too

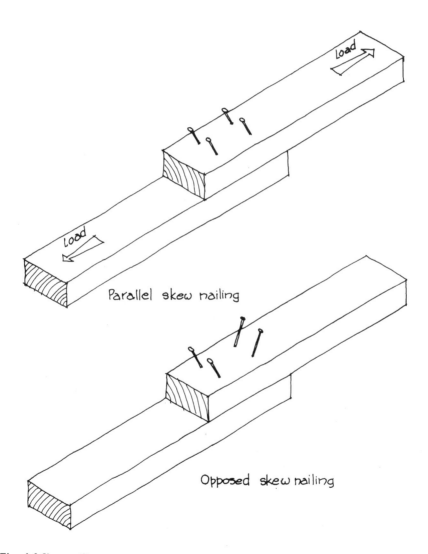

Fig. 4.6 Skew nailing

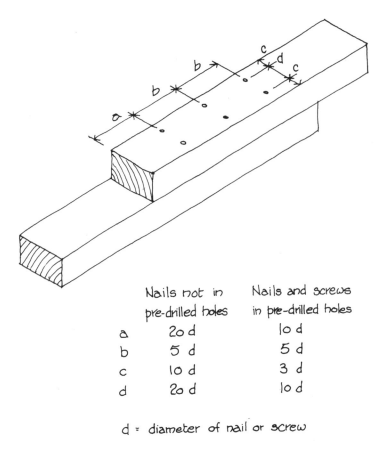

	Nails not in pre-drilled holes	Nails and screws in pre-drilled holes
a	20 d	10 d
b	5 d	5 d
c	10 d	3 d
d	20 d	10 d

d = diameter of nail or screw

Fig. 4.7 Spacing of nails and screws

close together their performance will be adversely affected. The minimum spacings recommended are shown in Fig. 4.7.

Screws, the British Standard states, should be screwed into pre-drilled holes not hammered in! The shank hole or clearance hole should be equal to the diameter of the shank and a pilot hole should be equal to about half the diameter of a shank.

The recommendations of the Standard for bolted joints are based on the use of black bolts to BS 4190 and washers complying with BS 4320. The diameter of the bolt is based on the stresses permitted and its length should be such as to leave one complete thread protruding from the nut when tightened. Washers should be provided under both the bolt head and the nut. They should be three times the bolt diameter across and of a thickness equal to 0.25 times the bolt diameter. If the washer is square, the thickness is the same as for round washers and the side length should be equal to the diameter.

Table 4.3 End and edge distances for nails and screws

Load direction	Distance	Nails not in pre-drilled holes	Nails or screws in pre-drilled holes
Parallel to grain	End distance	20 d	10 d
	Distance between adjacent nails or screws in any one line	20 d	10 d
Perpendicular to grain	End distance	5 d	5 d
	Distance between lines of nails	10 d	3 d

(*Source*: Based on BS 5268: *Structural use of timber*)

The bolt holes should be as near as possible to the bolt diameter and not more than 2 mm greater. They should be set away from the edge of the timber by not less than 1.5 times the diameter of the bolt and spaced not less than four times the diameter (see Fig. 4.8) except where the bearing is perpendicular to the grain. In this case, the edge distance should be four times the diameter and the spacing reduced to three times the diameter of the bolt if the ratio of member thickness to bolt diameter is one or five times where the ratio is three or more. Intermediate values for ratios between one and three can be interpolated.

The distance from the end of the timber member to the bolt hole should be four times the bolt diameter where the bearing direction is away from the end and seven times where it is towards it.

Three types of timber connector are covered in BS 5268: toothed plate, split ring and shear plate connectors, see Fig. 4.9. All these are fitted round a bolt to increase the strength of the joint. The recommended method for inserting toothed plate connectors is to fit the timber members together, clamp and drill the bolt hole right through. The clamps are removed and a high tensile steel threaded rod is passed through the holes in the timbers with the connector fitted round it. Steel plate washers, larger in diameter than the connector, are fitted to both sides, nuts are run down the rod and the whole assemblage tightened up. When the connector is fully embedded, the joint is again clamped, the rod released and withdrawn and the permanent bolts inserted.

Split ring and shear plate connectors should be fitted by drilling the bolt hole as described and then grooving or recessing the contact faces round the hole with a special tool to take the connector. Alternatively, if a jig or template is used the groove or recess can be cut simultaneously with the bolt hole. Once the timber is prepared, the connector is placed in position, the timbers assembled, the bolt, with its washers, inserted and the joint tightened up.

Glued joints receive limited coverage in the Standard which states that box

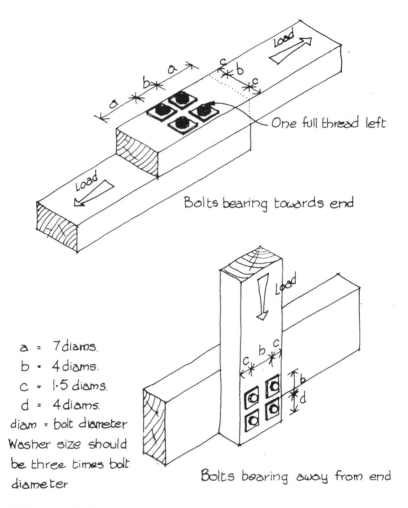

a = 7 diams.
b = 4 diams.
c = 1·5 diams.
d = 4 diams.
diam = bolt diameter
Washer size should
be three times bolt
diameter

Fig. 4.8 Spacing of bolts

beams, single web beams, stressed skin panels and glued gussets should be manufactured in accordance with BS 5291. Some guidance is given on permissible adhesives related to exposure and it is also pointed out that, when mechanical fasteners are present in a glued joint, they should not be considered as contributing to the strength of the joint.

Finger joints (see Fig. 4.7) are used to join two timbers end to end, but BS 5268 states that they should not be used in principal members in a frame or for any other member acting alone where the failure of the joint could lead to collapse. They may be used in other situations provided that it can be shown that if any one joint is removed, the remaining structure can continue to support a load equal to the full dead load of the structure plus any plant,

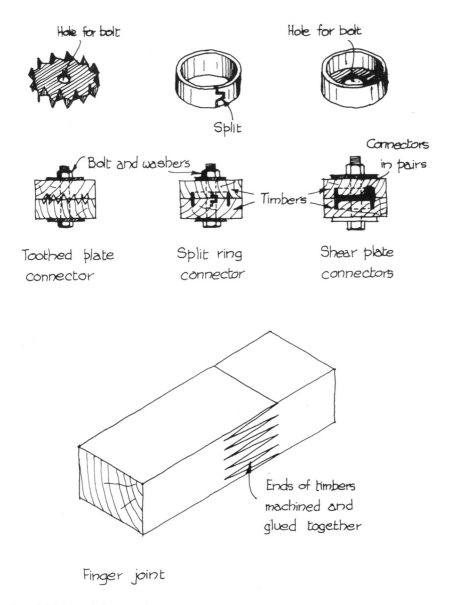

Hole for bolt

Hole for bolt

Split

Connectors in pairs

Bolt and washers

Timbers

Toothed plate connector

Split ring connector

Shear plate connectors

Ends of timbers machined and glued together

Finger joint

Fig. 4.9 Joining timber members

machinery or storage loads and one-third of any other imposed and wind loads. The new Eurocode 5, mentioned at the start of this section, will require that large finger joints or those in principal members must have third party quality assurance by a certification body that has been approved by the National Accreditation Council of Certification Bodies (NACCB).

4.4.3   Maintenance of timber structures

It is essential to maintain all vapour barriers, ventilation provisions and the like in an effective condition for the intended life of the structure and the design should ensure that access can be gained to those parts requiring periodic inspection and maintenance.

Bolts should be tightened after the first six to eight weeks of the building's life and, if the members are large and heavy, should be inspected again after twelve months. The bolts should also be tightened periodically if the moisture content of the timber fluctuates by more than 10 per cent.

4.5   Structural work of masonry

Unless the building is small enough to fit the specification given in Section 4.2, any external walls to be built of brick or similar materials must be constructed in accordance with the recommendations of BS 5628: Parts 1 and 3. In drafting these Standards the Committee assumed that the design of structural work in masonry would be entrusted to a chartered structural or civil engineer and, in practice this is the best policy since there are many complex factors to be taken into account. There are, however, a number of points in BS 5628 that are worth noting.

The term 'masonry' is defined in the Standard as an assemblage of structural units, either laid in-situ or prefabricated, which are solidly put together with mortar or grout.

The basis of design should ensure an adequate margin of safety and safety factors should be introduced to take account of:

- Possible unusual increases in load.
- Inaccurate assessment of the effects of the building loads.
- Unforeseen redistribution of stress within the structure.
- Variations in the dimensions due to construction tolerances and inaccuracies.

Consideration must also be given to the risk of adverse effects on the structure arising from expansion or contraction due to temperature or moisture changes, creep, settlement or deformation of flexural members. The Standard recommends the use of suitable detailing that will maintain an adequate margin of safety.

The aim should be to ensure a robust and stable building. For this one must examine the lay-out of the structure on plan, the returns at the ends of the walls, the interaction between intersecting walls and between masonry walls and other parts of the structure. In addition the building should be designed to withstand a horizontal uniformly distributed load that is equal to 1.5 per cent of the total design dead load. Tie straps should be provided at all points where the floors and roof join the walls except where the floor or roof is a concrete slab and it bears on the wall. Figure 4.10 shows some recommended strapping arrangements.

The dead load of the building, which is to be taken as the total weight of the structure plus its finishes and any partitions and fixtures, should be calculated in accordance with BS 6399: Part 1. The same Standard should be used to calculate the imposed loads. For wind loads the recommendations given in CP3: Chapter V: Part 2 should be followed.

The slenderness ratio of the walls is another consideration. This is the ratio of the effective height or effective length to the effective thickness and in most cases it should not exceed 27. The exemption is where the wall is in a two-storey building and is less than 90 mm thick, in which case the slenderness ratio should be limited to 20.

Effective height is taken as either:

- 0.75 times the height of the wall between storeys where the floors and roof are of *in-situ* or precast concrete construction and bear at least half way onto the thickness of the wall (or inner leaf if it is a cavity wall), subject to a minimum bearing of 90 mm, or
- the clear height between floors and roof where the construction is of the type shown in Fig. 4.10.

Effective length is measured as:

- 0.75 times the distance between intersecting walls properly bonded to the wall, or
- twice times the distance from such a bonded intersecting wall to a free edge, or
- the actual distance between intersecting walls where these are tied to the wall with metal anchors at 300 mm vertical spacing, or
- 2.5 times the distance from such a tied intersecting wall to a free edge.

4.5.1 Bricks

BS 5628 is concerned with the durability and use of bricks and refers to two other Standards for dimensional recommendations. The two Standards are BS 3921 for bricks with a nominal $225 \times 112.5 \times 75$ mm format and BS 6649 for those with a $200 \times 100 \times 75$ mm format. The latter have been specifically introduced for use in buildings requiring dimensionally co-ordinated sizes.

Included in the general recommendations is that the bricks should be laid with any frogs uppermost and the maximum height of wall that should be raised in a day is 1.5 m.

Durability is expressed in terms of frost resistance and soluble salt content. Three grades of frost resistance are given: those bricks that are fully frost resistant are designated F, those with moderate resistance are designated M and those which are not resistant are designated O. There are two grades of soluble salt content, low, designated L and normal, designated N. Both designation letters should be given, thus a brick described as FL is frost

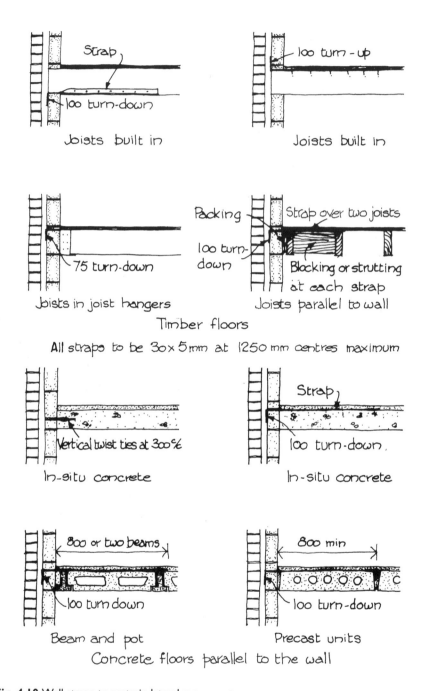

Fig. 4.10 Wall straps to restrain lateral movement

resistant with a low salt content and, therefore, would be very durable, whereas one given the designation ON would not be suitable for use in any situation where frost and moisture could affect it.

Frost-resistant bricks are those which are durable in all building situations, including those where they are in a saturated condition and subject to repeated freezing and thawing. Moderately resistant bricks are durable in all situations except that of repeated freezing and thawing when saturated and those which are not frost resistant are liable to damage unless protected during and after construction.

A low salt content would fall below the following:

Calcium	0.30%
Magnesium	0.03%
Potassium	0.03%
Sodium	0.03%
Sulphate	0.50%

A normal concentration is anything above those limits, but high concentration of salt in bricks is excluded by a clause in the Standard restricting the amount of efflorescence deposited on the face under test to not more than 50 per cent of the face area and no powdering or flaking.

4.5.2 Concrete blocks

Much use is made of concrete blocks in non-domestic work and BS 5628 makes a number of recommendations regarding their use.

In making a selection from the wide range of types of concrete block available, consideration should be given to durability, strength, adhesion, fire resistance, thermal and acoustic properties, handling, including the weight involved and appearance.

The durability of concrete blocks generally is high. They possess good frost resistance and most types are able to withstand the effects of repeated wetting or wetting and freezing. Some aerated blocks, however, may not be suitable in situations where there is a possibility that the blockwork will become saturated. The manufacturer's advice on this should be sought.

The strength of concrete blocks varies as widely as the range of blocks on the market. Aerated blocks tend to be those with the lowest strength and dense blocks have the highest. Nonetheless, even the lowest strength blocks are adequate for many of the applications found in buildings.

Suction can be a problem in the laying of some concrete blocks. The BS recommends that blocks should not be wetted to overcome this difficulty, instead the mortar mix should be adjusted to suit the rate of suction or else water-retaining admixtures can be used if necessary.

All concrete blocks possess a high degree of fire resistance and can be used with safety.

Thermal and acoustic properties are referred to in other chapters of this book, but generally vary in proportion to the weight, the heavier blocks being better sound absorbers and insulators than the lighter blocks.

The handling aspects are largely concerned with the avoidance of damage that would affect the appearance and, although an important consideration in the constructing of the building, is not a subject for the Building Regulations.

4.5.3 Mortar

The selection of the materials to be used and their proportions in the mix of mortar are to match the bricks used, the building loads anticipated and the degree of exposure of the building. The recommendations of BS 5628 are shown in Table 4.4. It should also be noted that the way the mortar joint is finished will affect the resistance of the wall to rain penetration. A bucket handle or weathered joint will give the best resistance, followed in order by a flush struck joint, then a recessed tooled joint and, finally, the joint giving the poorest resistance to rain is one that is recessed and untooled.

Table 4.4   Mortar mixes

		Mortar designation	Type of mortar proportion by volume			Mean compressive strength at 28 days (N/mm^2)	
			Cement: lime:sand	Masonry cement: sand	Cement: sand with plasticizer	Preliminary (lab. test)	Site test
(i)			1:0–0.25:3	–	–	16.0	11.0
(ii)			1:0.5:4–4.5	1:2.5–3.5	1:3:4	6.5	4.5
(iii)			1:1:5–6	1:4–5	1:5–6	3.6	2.5
(iv)			1:2:8–9	1:5.5–6.5	1:7–8	1.5	1.0

Increasing strength attack during construction ↑

Increasing ability to accommodate movement due to settlement or temperature and moisture changes ↓

→ Increasing resistance to frost attack during construction

← Improvement in bond and consequent resistance to rain penetration

(Source: Based on BS 5628: Part 1: Structural use of unreinforced masonry)

4.5.4 Damp-proof courses

The recommendation of the Standard for the installation of damp-proof courses follows normal good building practice by specifying the following positions:

- Horizontally and vertically where the ground floor is below ground level.
- 150 mm above ground level.
- Under pervious or jointed cills or sub-cills – the d.p.c. should be turned up at the back if the cill is in contact with the backing wall.
- At jambs with the frame so placed that the water cannot penetrate past the vertical d.p.c.
- Over openings with the cavity tray continuous with the vertical d.p.c. and at least two weep holes left.
- In parapets at 150 mm above the roof abutment so arranged as to form a moisture-resistant continuity with the roofing.
- Under copings.

4.5.5 Cavity walls

If the walls are of cavity construction each leaf is to be at least 75 mm thick and the cavity can be between 50 and 150 mm wide; however, it should not exceed 75 mm wide where either leaf of the wall is less than 90 mm thick.

It is also recommended that if the building is over four storeys or 12 m high, the outer leaf of the external walls should be supported at every third storey or at vertical intervals of 9 m.

Wall ties are to be provided of the types shown in Table 4.5 and illustrated in Fig. 4.11 at the rate of 2.5 ties/m², increasing to 4.9 ties/m² where either leaf is less than 90 mm thick. This rate is generally achieved with ties at 450 mm vertical spacing and 900 mm horizontal spacing, closing up to 750 mm horizontally for the thinner leaves. Extra ties are required at window reveals, vertical unreturned edges, movement joints and up the line of sloping verges spaced at 300 mm along a line not more than 225 mm from the edge. The length of the ties should be sufficient to ensure that at least 50 mm is embedded in the masonry joint at each end.

4.5.6 Movement joints

When the panels of brickwork are large, movement joints should be incorporated to accommodate both expansion and contraction. They should be built in as the work proceeds, not cut in afterwards.

Generally, the size of a brick panel should not exceed 15 m without an expansion joint and the width of the joint in millimetres should be approxi-

Table 4.5 Wall tie types and lengths

Least leaf thickness (one or both leaves) (mm)	Nominal cavity width (mm)	Permissible type of tie	Tie length (mm)
75	75 or less	Butterfly Double triangle Vertical twist	175
90	75 or less	Butterfly Double triangle Vertical twist	200
90	76–100	Double triangle Vertical twist	225
90	101–125	Vertical twist	250
90	126–150	Vertical twist	275

(*Source:* Based on BS 5628: Part 1: *Structural use of unreinforced masonry*)

mately 30 per cent more than the spacing of the joints in metres, i.e. with, say, a 12 m spacing the joint width in millimetres would be $12 + 30\% = 16$ mm.

4.5.7 Accidental damage

Not only is BS 5628 concerned with the manner of designing and building walls, it also makes recommendations about provisions to resist the effects of accidental damage. In buildings up to four storeys, the recommendation is that the plan form and construction should be sufficiently robust to contain the spread of damage should it occur. Where the building is of five storeys and over, the structure should not only be robust but also follow one of three additional recommended options.

Option 1 is that the vertical and horizontal elements, unless protected, can be shown to be removable, one at a time, without causing collapse. By 'protected member' the BS means that it can support the normal design loads plus an accidental design load of $34\,kN/m^2$ applied in any direction. The Standard points out that a masonry wall or column may have enough strength to withstand this pressure if it supports a sufficiently high vertical axial load.

Option 2 is to use horizontal ties to perimeter and internal walls and columns where the vertical elements, unless protected, can be shown to be removable, one at a time, without causing collapse. Ties should be fixed at each floor and roof level unless the roof is lightweight, i.e. timber or steel trusses, a timber flat roof or a roof with concrete or steel purlins and woodwool, or similar, slabs.

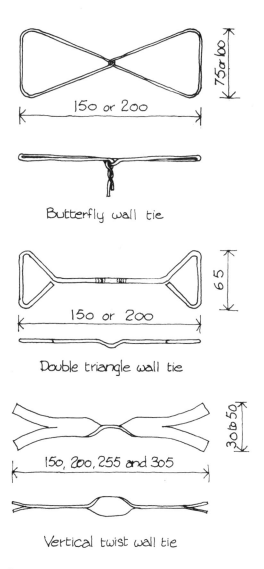

Fig. 4.11 Cavity wall ties

Option 3 is to have both horizontal and vertical ties as set out in the Standard.

4.6 Structural work of concrete

The Building Regulations refer to British Standard 8110 for the provision that would be deemed to satisfy the requirements. BS 8110 covers both reinforced and prestressed concrete structures and assumes the work will be designed and supervised by a qualified structural or civil engineer. It sets out parameters for use in one of three design methods: the elastic or permissible stress method; the ultimate load method; and the limit state method.

The elastic or permissible stress method aims at ensuring that the working stresses do not exceed a set of permissible stresses obtained by finding the actual strengths of the materials and dividing them by a factor of safety. The working stresses are taken as the worst conditions induced in the structure by all the working loads that can reasonably be expected to occur in practice.

The ultimate load or load factor method aims to ensure that the strength of the structure calculated from the actual strengths of the materials is sufficient to support an ultimate loading. This ultimate loading is found by multiplying the working stresses by a factor of safety.

These two methods appear to be identical but are only so for materials that are fully elastic right up to the point of failure. They can, however, be made to give very similar results but having two methods in one Standard (as it was in CP 114) has led to confusion in the minds of designers as to what their calculations were actually predicting.

The third method, the limit state, was introduced in BS 110 in 1972 and develops the logic of the load factor design method further. Instead of an allowance for all the uncertainties being covered by one factor of safety, the method defines a set of partial safety factors, one for each material used and for each type of load. This method avoids the weakness that is inherent in the other two when considering a structure subject to two simultaneous loads such as a vertical dead load and a wind load. In these circumstances, a critical design condition arises when one is at a maximum and the other is at a minimum.

Analysis of the structure has a fundamental aim which is to obtain a set of internal forces and moments throughout the structural frame that are in equilibrium with the design loads calculated from the required loading conditions. This analysis is the task that is set to the Structural Engineer and is beyond the scope of this book.

British Standard 8110 will be replaced in the near future by Structural Eurocode 2, produced by the European Standards Organization CEN. At the moment the Code is available to designers on a trial basis. Structures designed in accordance with the principles contained in EC2 would not conform to the recommendations contained in Approved Document A, however, it is possible that a local authority may accept such calculations since, ultimately, the Code will be the source of reference in the Approved Document.

4.6.1   Materials

The Standard refers to other British Standards for the specification of materials as follows:

Cement:	BS 12	for ordinary Portland cement and rapid hardening Portland cement
	BS 1370	for low heat Portland cement
	BS 40277	for sulphate-resistant Portland cement.
Aggregates:	BS 877	for foamed or expanded blast furnace slag
	BS 882	for aggregates from natural sources
	BS 1047	for air-cooled blast furnace slag
	BS 3797	for a specification for lightweight aggregates.

The maximum size of aggregate should be not more than one-quarter of the minimum thickness of the concrete section. Generally, a nominal size of 20 mm is suitable but, where there are no restrictions to the flow, 40 mm nominal may be used. In thin sections or where the reinforcement is closely spaced, a maximum size of 14 mm or even 10 mm should be considered.

Normal weight concrete is made with aggregate of a particle density between 2000 and 3000 kg/m^3. Aggregates less than this density produce lightweight concrete and those of greater density are used where heavyweight concrete is needed.

4.6.2 Durability

Not only must the structure support the loads to which it is subjected, it must do so for as long as the building is needed. One of the main characteristics affecting the durability of a concrete frame is its permeability to the ingress of water, oxygen, carbon dioxide and other potentially deleterious substances. This permeability is governed by the constituents of the concrete and the procedures employed in the making.

The other characteristics to be considered in relation to durability are:

- The shape and bulk of the concrete.
- The cover to the embedded steel.
- The type of environment, particularly the degree of exposure.
- The type of cement used.
- The type of aggregate used.
- The cement content and water/cement ratio.
- The workmanship employed.

The shape of the concrete section should be designed to promote good drainage of water from the surfaces and to minimize cracks. Consideration should also be given to increasing the cover to the steel at the corners, chamfering the corners, using circular cross-sections and treating the surfaces with a suitable coating.

Workmanship should be such as will obtain full compaction of the concrete and efficient curing.

4.6.3   Cover for reinforcement

The size of the concrete members must be sufficient, not only to accommodate the stresses analysed by the engineer but, in addition, must provide an adequate thickness outside the line of the reinforcement to provide the steel with a degree of protection appropriate to the conditions.

BS 8110 states that the cover for reinforcement should:

- comply with the recommendations related to bar size, aggregate size and for concrete cast against an uneven surface
- protect the steel against corrosion
- protect the steel against fire
- allow for any surface treatment such as bush hammering

The recommendation concerning bar size is that the thickness of the cover should be not less than the size of the main bars, or where the bars are in pairs or bundles, it should be not less than the size of a single bar equal to the sum of their cross-sectional areas and the bar size should be not less than the nominal size of aggregate used.

Table 4.6 shows the cover thicknesses required for differing degrees of exposure and grade of concrete. If the concrete is cast against an uneven surface, these thicknesses should be increased to ensure that the minimum cover achieved complies with the recommendations. Where the uneven surface is earth the minimum cover thickness should be not less than 75 mm and where it is blinding it should be not less than 40 mm.

The cover necessary to achieve specified periods of fire resistance is shown in Table 4.7 and the cover required where the concrete surface is to receive further treatment after casting should be sufficient so that at least the minimum recommended is left.

4.6.4   Transport, placing and compacting

The freshly mixed concrete should be moved as quickly as possible to its intended position in a manner that maintains its workability, prevents the segregation or loss of any of the constituents, and protects against the ingress of foreign matter or water. If a delay in placing the concrete occurs it can still be used provided that it can be worked and compacted without adding any more water.

The mix should be placed so as to avoid displacing any of the reinforcement, or displacing or damaging any of the formwork.

Compaction should be by vibration or other means applied continuously during placing to ensure that the concrete is thoroughly worked round the reinforcement and into the corners of the formwork. It should continue until all entrapped air has been removed. Excessive vibration can lead to a weak surface layer and should be avoided.

Table 4.6   Concrete cover required to protect reinforcement

Grade of concrete					
Lowest grade of concrete	C30	C35	C40	C45	C50
Maximum free water/cement ratio	0.65	0.60	0.55	0.50	0.45
Minimum cement content (kg/m³)	275	300	325	350	400
Conditions of exposure of surface	*Nominal cover* (mm)				
Protected against weather or aggressive conditions	25	20	20	20	20
Sheltered from severe rain or freezing while wet, subject to condensation, continuously under water, or in contact with aggressive soil	–	35	30	25	20
Exposed to severe rain, alternate wetting and drying, occasional freezing, or severe condensation	–	–	40	30	25
Exposed to sea water spray, de-icing salts or corrosive fumes, or severe freezing while wet	–	–	50	40	30
Exposed to abrasive action, flowing water with pH 4.5 or over, machinery or vehicles	–	–	–	60	50

(*Source*: Based on BS 8110: *Structural use of concrete*: Part 1: *Code of practice for design and construction*)

4.6.5   Curing

The curing of the concrete should start immediately after it has been compacted and is intended to prevent:

- premature drying out by the sun or wind
- leaching out by rain
- rapid cooling
- high internal temperature differences
- damage by low temperatures or frost
- disruption of the bond between the steel and the concrete by vibration or impact

The minimum periods of curing depend on the type of cement used and the environmental conditions. A dry situation – where the relative humidity is less than 50 per cent and the surfaces are totally unprotected from the sun and wind – would require between 4 and 10 days curing. With a humidity

Table 4.7 Concrete cover and beam and column widths required for fire resistance

Fire-resistance period (mins)	Minimum nominal cover (mm)			Minimum width (mm)			
	Beams		Columns	Beams	Columns		
	Simply supported	Continuous			Fully exposed	50% exposed	One face exposed
30	20	20	20	200	150	125	100
60	20	20	20	200	200	160	120
90	20	20	20	200	250	200	140
120	40	30	25	200	300	200	160

(*Source:* Based on BS 8110: *Structural use of concrete: Part 1: Code of practice for design and construction*)

between 50 and 80 per cent and some protection the time should be between 3 and 6 days but where the concrete is in a position where the humidity is over 80 per cent and the surface fully protected, no special provisions are required.

Three methods of curing are described in BS 8110, they are:

- To keep the formwork in place.
- To cover the surface with polythene sheet or a similar material, well sealed and fastened.
- To spray on a curing membrane.

4.6.6   Concreting in cold weather

Cold weather working of concrete is permitted provided a number of precautions are observed. These are intended to keep the temperature of the mix up until it has had a chance to set. The main points are that the aggregate and all surfaces with which the concrete will come into contact should be free from snow, ice or frost and no part of the concrete should be below 5 °C at the time of placing nor be allowed to fall below this temperature. If water spray is to be used in the curing process it should be delayed until the strength of the concrete has reached $5\,\text{N/mm}^2$.

A satisfactory temperature can be maintained by:

- heating the water or the aggregate
- increasing the cement content
- incorporating an accelerator
- covering the faces of the formwork with insulation

- heating the formwork
- working within a heated enclosure

4.6.7 Formwork

The design of the formwork must ensure that it is strong enough to support the worst combination of loads that could be placed upon it. These arise through the self-weight of the formwork, the weight of the wet concrete and its reinforcement, the pressure of the wet concrete on the sides of the formwork, any constructional and wind loads plus any dynamic effects due to the placing, vibrating and compacting of the concrete.

Striking of the formwork should be delayed until the structure is up to design strength and properly cured. The Standard recommends that the times given in the tables in Report 67 produced by the Construction Industry Research and Information Association (CIRIA) should be followed.

4.7 Structural work of steel

Once again, the Building Regulations refer to the British Standards for the requirements. In this case the standard is BS 5950: *Structural use of steelwork in buildings*. In common with similar Standards, BS 5950 assumes that the execution of its provisions will be entrusted to appropriately qualified and experienced people.

Reference is made in Sections 4.4 and 4.6 to European Structural Codes. The design of steelwork, in common with that of timber and concrete, will in due course be based on the recommendations contained in these new standards. In the case of steel, the standard is the Structural Eurocode 3. This, like the Code for work in concrete, has been issued on a trial basis prior to its adoption in the not too distant future.

To follow the recommendations of BS 5950, steelwork should be designed using the limit states at which the members become unfit for their intended use. These limit states are found by applying appropriate factors to the ultimate limit states and the serviceability limit states. The ultimate limit states relate to the strength of the frame, the stability against overturning or swaying, fracture due to fatigue and brittle fracture. The serviceability limit states relate to deflection of the frame, vibration such as wind-induced oscillation, repairable damage due to fatigue, corrosion and durability.

The steel frame may be of simple design in which all the connections are bolted or riveted and considered to be pin joints that can rotate, rigid design in which the connections are welded and are stiff enough to develop a full continuity of the members in the frame, or semi-rigid design in which some stiffness is taken into account.

The steel, generally, should be in accordance with BS 4360; however, other steels may be used provided that due allowance is made for their differing qualities.

Welding consumables should conform to the recommendations of BS 5135 and bolts, nuts and washers should be to BS 4190 or BS 3692.

Bolts and rivets should be spaced not less than 2.5 times the nominal diameter of the bolt or rivet and not more than 14 times the least thickness of the metals being joined. End and edge distances should be 1.25 times the diameter of the fixing hole where the end or edge has been rolled, machine flame cut, sawn or planed and 1.4 times the diameter where the end or edge has been sheared or flame cut by hand.

The size of columns and beams must be calculated on the basis of the anticipated loading but the Standard lays down the minimum sizes that should be used to restrict deflection, these are:

- Cantilevers   length divided by 180
- Beams carrying plaster   length divided by 360
- Other beams   length divided by 200
- Purlins and sheeting rails   as required to suit the cladding system
- Columns   storey height divided by 300

4.8   Structural work of aluminium

The British Standard reference for this work is BS 118: *The structural use of aluminium.*

The aluminium used in structural frames is always an alloy of the metal and, as is normal, there are many differing alloys available. To distinguish one from another a system of nomenclature is used giving a code indicative of the material.

The first letter of the code is either H meaning that it can be heat treated, or N meaning that it cannot. Casting alloys are prefixed LM. Where relevant, there is a second letter indicating the form of product which is E for extruded sections like bars and tubes, T for tubes that have been drawn and S for sheet and plate products. This is followed by a number denoting the actual alloy; numbers between 3 and 8 indicate a non-heat-treatable metal and 9 onwards is given to metals that can be heat treated. Finally, for heat-treatable alloys there is a hyphen and a symbol to indicate the temper of the metal: O is annealed and the softest; M is as manufactured and so is partly hardened; H2 to H8 show progressive degrees of hardness; and TB, TF and TE indicate heat treatment processes.

The normal choice of alloy for bolted and riveted framed structures is H30-TF and N8 for welded frames. This latter alloy possesses exceptional durability, especially in marine environments. Where appearance is important, H9 can be used. This is of moderate strength but has a good surface that responds well to anodizing.

Most sections are produced by extrusion and, since the cost of dies is low, there are many varieties available. BS 1161 lists 60.

4.8.1 Design

The design procedure is, basically, the same as for a steel frame except that the high coefficient of expansion of aluminium must be borne in mind in the detailing and assessment should be made of the possibility of failure through fatigue. It is also recommended that the members and the connections are left readily accessible for maintenance purposes and designed to avoid any pockets or crevices where water, condensation or dirt can collect.

As with steel, the Standard gives minimum sizes to limit deflection:

- Beams carrying plaster   span divided by 360
- Purlins and sheeting rails   span divided by 200 (dead load only)
                                          span divided by 100 (worst combination of load)
- Curtain wall mullions   span divided by 175

Bolts and rivets should be spaced not less than 2.4 times the diameter of the bolt or rivet and not more than 16 times the least thickness of the metals being joined.

Welded joints can be used but allowance must be made for the fact that the weld can reduce the strength of the alloy in the vicinity of the joint.

Joints can also be glued using a thermo-setting adhesive but these should not be joints subject to tension nor those where the loading tends to force the joint apart.

4.8.2 Protection

Protection is often not required as aluminium is resistant to corrosion but it does depend on the alloy used and the environment in which it is placed. If the environment is mild, no protection is needed but if it is in an industrial area, some may be required to prevent the metal becoming dark and rough. If it is an aggressive environment, protection should be provided and, for some alloys, it is necessary in a marine location. Special precautions are needed when the alloy is to be immersed in water.

Special protection is also required where the aluminium is in contact with other metals or certain substances likely to react with it. The most severe damage results from water draining off copper or copper alloy surfaces, and this should be avoided by the design.

Protection can be provided by painting or by spraying on a coat of commercial purity aluminium.

4.9 Air-supported structures

Air-supported structures (i.e. space-enclosing single skin membranes, anchored to the ground and kept in tension by internal air pressure so that they can support a load) are sometimes used for assembly purposes, particularly of a sporting nature, but they are neither covered in the requirements laid down in the Approved Documents, nor is there any reference to a British Standard that should be followed. Nonetheless, as a building by definition, any such structures must comply with the fundamental requirements applied both to the structure and to other matters such as means of escape.

Although an air-supported structure is a complete enclosing envelope and includes the roof, it is dealt with here as it is impossible to separate the functions of wall and roof in this form of building.

The accepted structural guide for this purpose is BS 6661: *Design, construction and maintenance of single skin air-supported structures*. The following notes are all based on this publication.

4.9.1 Generally

The structure should be designed so that:

- It can safely sustain all forces imposed on it.
- It will perform adequately throughout its design life, taking into account any deterioration that could take place in the membrane and its joints.
- The materials possess adequate resistance to biological and chemical attack, weathering and abrasion.
- It possesses an appropriate level of safety with regard to the consequences of failure.
- There will be sufficient time for the occupants to escape in the event of collapse due to membrane damage or fan failure.

Protection against the problems arising out of deflation due to fan failure is covered in the Standard by the recommendation that in all buildings used for assembly purposes where the planned occupancy is greater than 25, there should be an automatically operated standby fan and alternative power source that will be activated in the event of either a pressure loss in the structure or a failure of the primary power source.

4..9.2 Design

The design of air-supported structures is a matter for expert attention but there are a number of points worth noting in the British Standard.

There are, at present, no satisfactory standard tests that can be applied to the membranes from which air-supported structures are created. The Standard, therefore, recommends that the relatively high factors of safety as set out should be adopted.

The material used for the membrane can be either a coated fabric or a plastic film. Since the majority of failures have arisen out of the tearing of the membrane, the tear strength is an important consideration and should be such that a failure will not occur if the material is damaged under a condition of maximum working load.

The full range of properties the material should possess is:

- tensile strength
- tear resistance
- resistance to burning
- resistance to elongation
- resistance to cracking due to flexing or cold conditions
- high resistance to water and air penetration
- resistance to ultra-violet light degradation
- resistance to fungal and chemical attack
- good coating adhesion (coated fabrics)

Coated fabrics are usually made from a woven base of nylon, polyester, polypropylene or glass fibre and coated with polyvinyl chloride, chloroprene rubber or chlorosulphonated polyethylene elastomer. The plastic films are usually either polyvinyl chloride or polyethylene sheeting. All the coated fabrics can be joined by stitching or by the use of adhesives, in addition, the PVC-coated fabrics can be joined by radio-frequency or dielectric welding. The plastic films should be joined by welding or by the use of adhesives.

The anchorage system can be by means of screw anchors, steel stakes, concrete block dead-load anchors or *in-situ* concrete foundations. Weather protection to the fittings must be provided and the connections should be readily accessible for inspection.

No recommendations are given in BS 6661 as to the type of fan needed to provide the air supply but its capacity should allow for all the air pressure losses that can occur, it should be fitted with a back-draught damper and able to function irrespective of the internal air pressure. Control of the fan should ensure that the pressure does not exceed the design inflation pressure by more than 25 per cent nor fall more than 85 per cent below it. Emergency fans, when required for safety reasons, should have a capacity that will maintain the required pressure in the structure and an alternative power supply that will come into operation within 30 seconds of the failure of the main power supply. The alternative power supply must be capable of supporting the operation of the emergency fans for at least one hour plus any time that the structure may be left unattended. All fan equipment should be located outside in a weather-protecting structure and positioned so that any diesel or other toxic fumes are directed away from the air-supported structure. Precautions should also be taken to avoid the effects of debris collecting in the fans, the icing of any grills and extreme wind conditions. The emergency fans should be isolated from the main fans to prevent them being damaged in the event of fire in the latter.

Revolving entrance doors or air locks should be provided to preserve the designed air pressure.

4.9.3 Maintenance

Because of the relatively lightweight nature and short life of air-supported structures, their correct operation and maintenance are of great importance. The manufacturer should supply a 'log book' containing details of the full design data and the operating and maintenance instructions. This 'log book' should be in the custody of a site controller who will be responsible for the correct operating and maintenance of the structure.

4.10 Wall cladding

Any framed structure requires to be clad to complete the building enclosure. Section 2 of Approved Document A provides details of the support and fixings that are considered would satisfy the requirements of the Regulations regarding safety, particularly with respect to the heavier claddings such as precast concrete panels. Section 5 of Approved Document C4 sets out the principles to be observed to ensure weather resistance.

The function of cladding is to resist the penetration of rain and snow to the inside of the building and consideration must be given to the effect of wind either driving the rain or snow into the joints in the cladding or forcing it off its fixings. It must be capable of safely sustaining and transmitting to the supporting structure of the building, all dead, imposed and wind loads.

Two systems of cladding are used. The first is an impervious skin, either jointless with sealed joints, attached to the face of the frame or building through which the rain cannot penetrate. The second is a weather-resisting layer with dry joints that are either designed to collect any water ingress and direct it back out to the face or backed by a moisture-resisting layer, see Fig. 4.12. This second system can accept structural or thermal movement more readily than the first.

Three types of cladding are listed in the Approved Document:

- Impervious   metal, glass, plastics and bituminous products.
- Weather resisting   natural stone, cement-based products, fired clay and wood.
- Moisture resisting bituminous and plastics products with overlapping joints, permeable to water vapour.

Moisture-resisting claddings and most of the weather-resisting claddings are permeable to water vapour and, therefore, a vapour barrier or ventilated cavity must be incorporated behind the cladding.

The loadings to be sustained should be derived from CP3: Chapter V: Part 2 using a Class A building size for determining the ground roughness factor.

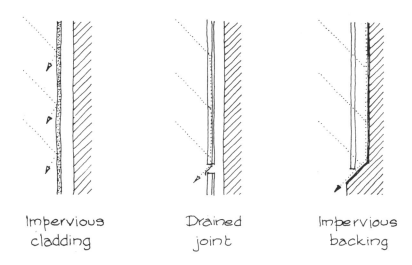

Impervious Drained Impervious
cladding   joint   backing

Fig. 4.12 Types of cladding

Allowance must also be made for any loads or forces due to ladders, access cradles or fixtures such as signboards, antennae and the like.

4.10.1 Fixings for wall cladding

The strength of the fixings should be derived from tests using materials representative of that into which the fixing will be made. BS 5080: Part 1 and Part 2 and the British Board of Agrément MOAT No. 19: *The assessment of torque expanded anchor bolts when used in dense aggregate concrete* can be employed as standards and references for the purpose of these tests.

Where expanding bolt type fixings are used, their assumed safe working shear and tensile strength values should be taken as the lower of either:

- one-third of the sum of the mean shear or tensile test load minus three times the standard deviation derived from the tests, or
- the mean of the loads that cause a displacement of 0.1 mm under direct tension and 0.1 mm under direct shear.

Where resin-bonded fixings are used, their assumed safe working shear and tensile strength values should be taken as the first of the above, that is, one-third of the sum of the mean shear or tensile failure test load minus three times the standard deviation derived from the tests. Account must also be taken of the rapid loss of strength that can occur in resin bonds at temperatures above 50 °C.

The component parts of mechanical fixings must be efficiently locked together so that any unintended slippage between the parts is prevented and designed to allow for an eccentricity in the application of the imposed load on the fixings. The allowance for the amount of the eccentricity should be 0.5 times the diameter of the fixing to avoid the possibility of local spalling.

The fixings must also allow for differential movement of the wall cladding and the supporting structure. Guidance on this is given in BS 8200 and BS 5628.

4.11 Thermal insulation

The Building Regulation L1 states that 'Reasonable provision shall be made for the conservation of fuel and power in buildings by limiting the heat loss through the fabric of the building.' Section 2 of Approved Document L (1995 edition) states that this requirement will be satisfied in relation to external walls if they possess a U value of $0.45 \, W/m^2 \, K$. U value is explained in Chapter 1.

There are three methods for showing the thermal efficiency of a building:

- The Elemental Method.
- The Calculation Method.
- The Energy Use Method.

Further information is given about these methods and how they are employed in Chapter 9 in relation to window and door openings.

To arrive at the thermal insulation value of a wall, Approved Document L shows, in Appendix A, the base thickness of thermal insulation required to give $0.45 \, W/m^2 \, K$ and the amount this base thickness can be reduced by allowances for the insulation value of the other components in the wall. This information is summarized in Table 4.8.

To use Table 4.8, first select the insulation material to be used in the wall and below it read off the required base thickness. Continue reading down the column to find the reduction allowances for the wall components, adjust each one to suit the actual thickness (if different from that given), and subtract it from the base thickness. The final sum will be the minimum thickness required. Typical constructions and their assessments are given in Fig. 4.13.

The values of thermal insulation given in Table 4.10 for timber frames and glass fibre quilt or slab are based on the value of the nominal 50×100 timbers combined with the insulating material.

The insulation shown in Fig. 4.13 as being necessary in a steel-framed and steel-clad wall is based on thermal insulation requirements. This may need to be increased to provide a satisfactory standard of insulation against radiant heat in the event of a fire, depending on the details of the construction and the location of the wall.

4.11.1 Thermal bridging

The procedure described in the last section allows for the variation that can occur in the thermal value of a construction due to different materials being

Table 4.8   Thickness of wall insulation for a design *U* values of 0.45 and 0.6 W/m² K

Criteria	Insulation material			
	Phenolic foam	Polyurethane board	Expanded polystyrene slab, glass fibre slab, or mineral fibre slab	Glass fibre quilt, or urea–formaldehyde foam
Thermal conductivity (W/m K)	0.020	0.025	0.035	0.040
Design *U* value of wall (W/m² K)	0.45 0.6	0.45 0.6	0.45 0.6	0.45 0.6
Base thickness of insulation (mm)	41 30	51 37	71 52	82 59

Wall components	Allowable reductions (mm)			
Brick outer leaf	2	3	4	5
100 mm concrete outer leaf kg/m³				
Aerated 600	8	10	15	17
800	7	8	12	14
Lightweight 1000	6	7	10	11
Dense 1600	3	3	5	5
100 mm concrete inner leaf kg/m³				
Aerated 600	9	11	15	17
800	7	9	13	15
Lightweight 1000	6	8	11	12
Dense 1600	3	4	5	6
50 × 100 nom. timber frame and glass fibre slab	42	53	74	84
50 × 100 nom. timber frame and glass fibre quilt	38	48	67	77
Cavity (25 mm minimum)	4	5	6	7
13 mm plaster	1	1	1	1
13 mm lightweight plaster	2	2	3	3
10 mm plasterboard	1	2	2	3
13 mm plasterboard	2	2	3	3
Airspace behind plasterboard dry lining	2	3	4	4
9 mm sheathing ply	1	2	2	3
20 mm cement render	1	1	1	2
13 mm tile hanging	0	0	1	1

(*Source*: Based on Tables A5, A6, A7 and A8 of Appendix A of Approved Document L (1995 edition))

present in different parts of the structure. For instance, the thermal transmission through the mortar joints in a lightweight concrete block wall is greater than that through the blocks themselves.

If one of the alternative methods of demonstrating compliance is used, this

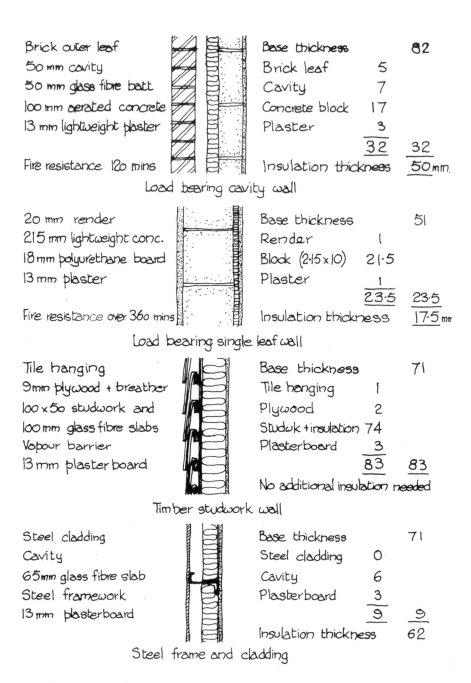

Brick outer leaf
50 mm cavity
50 mm glass fibre batt
100 mm aerated concrete
13 mm lightweight plaster

Fire resistance 120 mins

Base thickness		82
Brick leaf	5	
Cavity	7	
Concrete block	17	
Plaster	3	
	32	32
Insulation thickness		50mm

Load bearing cavity wall

20 mm render
215 mm lightweight conc.
18 mm polyurethane board
13 mm plaster

Fire resistance over 360 mins

Base thickness		51
Render	1	
Block (2·15 × 10)	21·5	
Plaster	1	
	23·5	23·5
Insulation thickness		17·5 mm

Load bearing single leaf wall

Tile hanging
9mm plywood + breather
100 × 50 studwork and
100 mm glass fibre slabs
Vapour barrier
13 mm plaster board

Base thickness		71
Tile hanging	1	
Plywood	2	
Studwk + insulation	74	
Plasterboard	3	
	83	83
No additional insulation needed		

Timber studwork wall

Steel cladding
Cavity
65mm glass fibre slab
Steel framework
13 mm plasterboard

Base thickness		71
Steel cladding	0	
Cavity	6	
Plasterboard	3	
	9	9
Insulation thickness		62

Steel frame and cladding

Fig. 4.13 External wall constructions

variation, or, as it is referred to in Approved Document L1, thermal bridging, must be entered in the calculations to obtain an average U value for the wall.

It is worth noting that, in this context, the term 'thermal bridging' has a slightly different connotation than when it is used in connection with a window or a door. In this case it refers to variations in the thermal value, whereas in the case of a window it refers to the heat loss route formed by the closure to the opening.

The Chartered Institute of Building Service Engineers (CIBSE) has published a Design Guide in which Section 3 gives a method for the calculation of thermal bridging variations in a wall. Examples from this Guide are given in Appendix B to Approved Document L1.

The method works by calculating the thermal resistance through each of the various materials present, taking the proportional area of each and combining them to give the total resistance of the wall. In a cavity wall, each leaf is calculated separately, including half the cavity resistance, and the two added together to give the total resistance, the reciprocal of this sum gives the U value.

The following example demonstrates the method:

Check that the wall illustrated at the top of Fig. 4.12 achieves the desired U value of $0.45\,\text{W/m}^2\,\text{K}$. Assume the normal thermal conductivity values of $0.84\,\text{W/m K}$ for the brickwork, $0.19\,\text{W/m K}$ for the blockwork, $0.8\,\text{W/m K}$ for the mortar, $0.16\,\text{W/m K}$ for the lightweight plaster and $0.04\,\text{W/m K}$ for the glass fibre insulating batts. Assume that the size of the concrete blocks is $440\,\text{mm} \times 215\,\text{mm}$ and that the mortar joints are $10\,\text{mm}$ thick (and, of course, $100\,\text{mm}$ wide).

Since the difference in thermal resistance between the mortar and the brick can be shown to be less than $0.1\,\text{m}^2\,\text{K/W}$ any bridging effect can be ignored and the outer leaf treated as one material.

Where a cavity is included in the structure, the value of the resistance of the cavity ($0.18\,\text{m}^2\,\text{K/W}$) is divided equally between the two leaves.

Material resistances
(thickness in metres divided by the thermal conductivity)

Brick	$\dfrac{0.102}{0.84} =$	$0.12\,\text{m}^2\,\text{K/W}$
Concrete blocks	$\dfrac{0.10}{0.19} =$	$0.52\,\text{m}^2\,\text{K/W}$
Mortar	$\dfrac{0.10}{0.80} =$	$0.125\,\text{m}^2\,\text{K/W}$
Plaster	$\dfrac{0.013}{0.16} =$	$0.08\,\text{m}^2\,\text{K/W}$
Insulation	$\dfrac{0.05}{0.04} =$	$1.25\,\text{m}^2\,\text{K/W}$

Surface resistances
Inner wall face   = $0.12\,\mathrm{m^2\,K/W}$
Outer wall face   = $0.06\,\mathrm{m^2\,K/W}$
Cavity   = $0.18\,\mathrm{m^2\,K/W}$ (half cavity $0.09\,\mathrm{m^2\,K/W}$)

Resistance of heat flow paths
Inner surface / plaster / concrete block / insulation / half cavity:
    $0.12 + 0.08 + 0.52 + 1.25 + 0.09 = 2.06\,\mathrm{m^2\,K/W}$
Inner surface / plaster / mortar joint / insulation / half cavity:
    $0.12 + 0.08 + 0.125 + 1.25 + 0.09 = 1.67\,\mathrm{m^2\,K/W}$

Fractional area of materials
Concrete blocks $\dfrac{440 \times 215}{450 \times 225}$ = 0.934

Mortar $1.000 - 0.934$ = 0.066

Total resistance of the inner leaf
Sum of parallel resistances through the concrete blocks and through the mortar:
    $(2.06 \times 0.934) + (1.67 \times 0.066) = 2.03\,\mathrm{m^2\,K/W}$ (R_i)

Total resistance of the outer leaf
Resistance of the outer surface   = $0.06\,\mathrm{m^2\,K/W}$
Resistance of the brickwork   = $0.12\,\mathrm{m^2\,K/W}$
resistance of half cavity   = $0.09\,\mathrm{m^2\,K/W}$
Total resistance of outer leaf   = $0.27\,\mathrm{m^2\,K/W}$ (R_o)

Total thermal resistance of the wall
The sum of the resistance of the two leaves, i.e.:
    $R_\mathrm{i} + R_\mathrm{o} = 2.03 + 0.27 = 2.30\,\mathrm{m^2\,K/W}$

U value of the wall
This is given by:

$$\frac{1}{(\text{total resistance})} = \frac{1}{2.30} = 0.434$$

Since 0.434 is less than the target figure of 0.45, the wall insulation standard is satisfactory. Further examples of calculations are given in Appendix B to Approved Document L.

4.12   Fire safety

The principle applied in the requirements for fire safety in external walls is that there should be sufficient fire resistance to prevent a spread of fire across the boundary. This may be the actual boundary of the site or a notional boundary drawn between two buildings. The chance of a fire spreading across an open space between buildings depends on:

- the size and intensity of the fire
- the distance between the buildings
- the fire protection given by their opposing faces

The size of the fire can be controlled by dividing the building into fire-resisting compartments (see Chapter 6 for fire-resisting floors and Chapter 8 for fire-resisting internal walls). The intensity of the fire is related to the use of the building and it can be moderated by the use of a sprinkler system.

Consideration must also be given to the risk presented to people in the other building and it is recognized in the Approved Documents that, unless the other building is used for residential, recreational or assembly purposes, the risk to life is low and can be discounted if it is on the same site.

One particular building use – hospitals – must receive special consideration. There is a form of hospital design known as the Nucleus plan. This consists of a number of standard cruciform units or 'templates' connected to each other and linked by a 'hospital street', see Fig. 4.14. Details of the standards to be achieved in this form of design are produced by NHS Estates. Each template must be designed to contain fire and smoke for at least 60 minutes so the perimeter walls of one such template, whether external or internal, must conform to this standard. The concept of these fire-containing units is intended to assist the difficult problem of evacuation and is dealt with in detail in Chapter 15.

In any building enclosure, no matter what the use, two aspects must receive attention: the fire resistance of the structural frame, if one is used, and the fire resistance of the wall structure, especially the external face. In the case of loadbearing masonry walls, the face and the structure are integral.

4.12.1 Structural frames

Any multi-storey structural frame, or 'element of structure', as it is termed in the Regulation, must possess a degree of fire resistance expressed in minutes. Single-storey frames that only support a roof are excluded from the requirements unless the roof is used as a means of escape or for car parking or the frame is essential for the stability of an external wall that needs to have a fire resistance. The frame in a building for public assembly or residential use should be capable of maintaining its loadbearing capacity in a fire for between 30 and 120 minutes, depending on the height of the building and the location of the frame (see Table 4.9).

The fire resistance of a reinforced concrete frame depends on the size of the member and the cover to the reinforcement (see Table 4.7). In the case of steel and aluminium structures, while neither metal is combustible or assists in the spread of fire, high temperatures lead to a serious loss of strength, particularly with aluminium which suffers a significant reduction at 250 °C

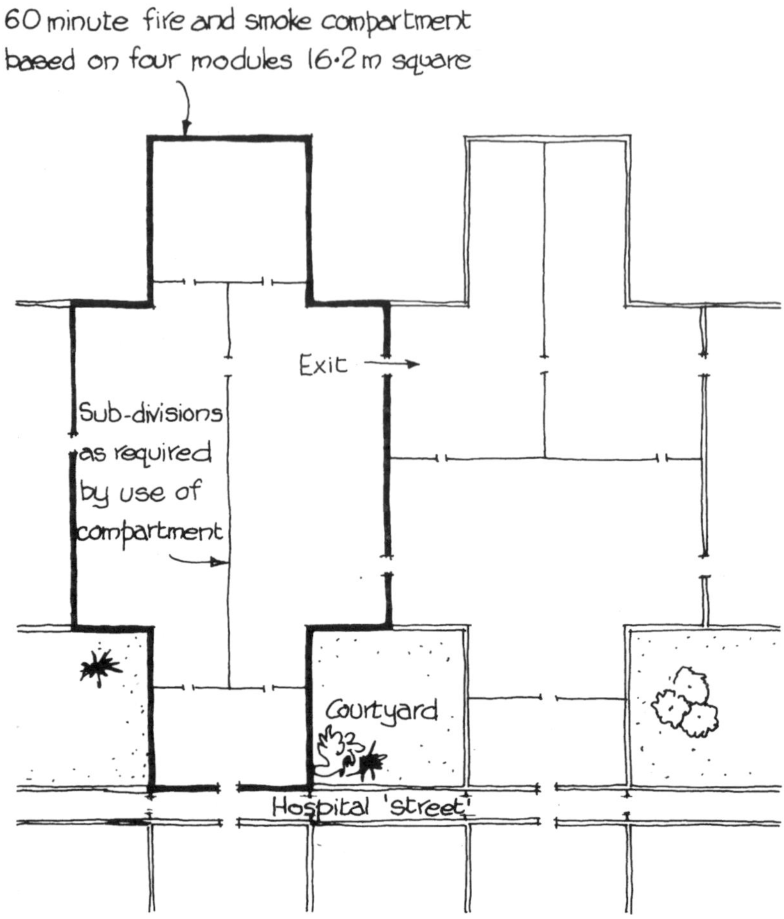

60 minute fire and smoke compartment
based on four modules 16·2m square

Exit →

Sub-divisions as required by use of compartment

Courtyard

Hospital 'street'

Fig. 4.14 Nucleus type of hospital plan

and melts at 650 °C. Both metals must be protected by suitable fire casings or encasement in a fire-resisting structure.

Portal frames, being the structure of the roof as well as the walls, present a special case. Normally, the frame would require no fire protection since it is usually in a single-storey building; however, it may be required to support an external wall that does need to offer some fire resistance because of its proximity to the boundary. In this case the column section of the portal frame must be fire resisting and, since the rafter is part of the same frame, it, too, must be protected otherwise the collapse of the rafter would also lead to a collapse of the wall.

Concrete portal frames possess an inherent fire resistance that would meet the requirements imposed. Steel frames do require protection throughout, unless the connection between the base of the frames and the foundation is designed as a rigid joint. In this case, the rafters can be left unprotected since their collapse in a fire would not result in the external wall failing to continue to perform its structural function. The design method to achieve this is set out in *Fire and steel construction: The behaviour of steel Portal frames in boundary conditions*, 1990, which is available from the Steel Construction Institute.

4.12.2 Wall constructions and cladding

Table 4.9 shows the fire-resistance standards required in external walls and Table 4.10 gives the notional fire-resistance of a selection of wall constructions. BS 5628 gives a more comprehensive list.

While the minimum thicknesses given might satisfy the fire-resistance requirements, it is possible that they may not be adequate either for the purposes of weather exclusion or for thermal insulation. For this reason, the constructions shown in Fig. 4.13 not only produce satisfactory thermal insulation standards and weather resistance but their fire-resistance capacity is also quoted.

Table 4.9 Minimum periods of fire resistance in frames and walls

Building purpose	Minimum periods of fire resistance (minutes)					
	Frames or walls in a basement storey		Frames or walls in the ground storey or an upper storey			
	Depth of the lowest basement (m)		Height of the top floor above ground level (m)			
	>10	<10	<5	<20	<30	>30
Public assembly						
Not sprinklered	90	60	60	60	90	NP
Sprinklered	60	60	30*	60	60	120†
Residential						
Any non-domestic use	90	60	30*	60	90	120†
Storage						
Not sprinklered	120	90	60	90	120	NP
Sprinklered	90	60	30*	60	90	120†

(*Source*: Based on Table A2 of Approved Document B to the Building Regulations 1991)

* Increased to a minimum of 60 minutes for compartment walls separating buildings.
† Reduced to 90 minutes for elements not forming part of the structural frame.
NP = Not permitted

Table 4.10   Notional fire resistance of loadbearing walls

Wall description	Finish (13 mm thick to both faces)	Minimum thickness of masonry (mm) for notional periods of fire resistance (min) of:			
		30	60	90	120
Single leaf walls					
Clay brick	None	90	90	100	100
	Vermiculite/gypsum	90	90	90	90
Perforated brick	None, sand/cement, sand/gypsum	100	170	170	170
	Vermiculite/gypsum	90	100	100	170
Concrete block with Class 1 aggregate	None	90	90	100	100
	Vermiculite/gypsum	90	90	90	90
Concrete block with Class 2 aggregate	None, sand/cement, sand/gypsum	90	90	100	100
	Vermiculite/gypsum	90	90	90	90
Aerated concrete	None	90	90	100	100
	Vermiculite/gypsum	90	90	90	100
Cavity walls – both skins					
Clay brick	None	90	90	100	100
Solid concrete	None	90	90	100	100
Aerated concrete	None	90	90	100	100

(*Source*: BS 5628: *Structural use of unreinforced masonry*: Part 3: Table 16)

Notes
1. Clay bricks are either without a frog or laid with frog up.
2. Class 1 aggregate is limestone, blastfurnace slag, crushed brick, clinker, pulverized fuel ash or pumice.
3. Class 2 aggregate is all gravels and crushed natural rock except limestone.

Wall surfaces and any cladding to a structural frame must, when facing a boundary, possess fire-resisting surfaces as shown below. A wall is considered to face a boundary if it is parallel to it or if it is at an angle that is less than 80°. If any cladding does not meet the standards given, its area must be restricted in size as shown in Fig. 4.15. Any thermal insulation incorporated within the cladding system must be of limited combustibility, which means that it must have a density of not less than 300 kg/m^3 and, when tested in accordance with BS 476: Part 11, does not flame and its rise in temperature while being tested does not exceed 20 °C. Where the system is of the type with a drained and ventilated cavity, the back surface of the cladding, facing the cavity, must meet the same requirements as the external surfaces, which are:

Distance of face from boundary	Height of face	Requirement
(Public assembly buildings)		
1 m or more	More than one storey	Index (I)[1] not more than 20[2] for first 10 m above the ground level or a roof or any part of the building to which the public have access. No requirements above 10 m
(Any building other than for public assembly use)		
1 m or more	Less than 20 m	None
1 m or more	More than 20 m	Index (I) not more than 20 for first 20 m of height and Class O[2] above 20 m
Less than 1 m	Any height	Class O[2]

[1] Index (I) is a fire propagation index determined by reference to BS 476: Part 6 in relation to materials that may ignite easily and refers to the overall test performance (there is also a sub-index (i_1) which refers to the first three minutes of the test).

[2] BS 476: Part 7 gives test methods to establish a rating for the spread of flame across a surface. These ratings are classified 1, 2, 3 or 4 with Class 1 being the highest. Class O is not found in any British Standard but is achieved by a material that is either of limited combustibility throughout or is a Class 1 material with a fire propagation index (I) of not more than 12 and a sub-index (i_1) of not more than 6.

4.12.3 Unprotected areas

Any part of an external wall that has less fire resistance than that shown in Table 4.9 must be considered to be an unprotected area. Small unprotected areas in what, otherwise, is a fully protective wall are not considered hazardous if they come within the limits shown in Fig. 4.15.

If an external wall, possessing the requisite fire resistance, is faced with a combustible material more than 1 mm thick an unprotected area is created but, since the backing is protective, the area to be taken for the purposes of defining the separation of buildings is counted as half the actual area.

There are two exemptions to the requirements of this Building Regulation on unprotected areas. The first is canopy structures. Most canopies are exempt from the Building Regulations under the rules set out in Schedule 3 but, it may be that due to its size or its distance from the boundary the canopy is not exempt. In this case it is recognized that the provisions about unprotected areas could be onerous and can be reasonably disregarded. The second exemption is any part of the external wall of an uncompartmented building that is more than 30 m above the ground may also be disregarded in assessing the unprotected areas.

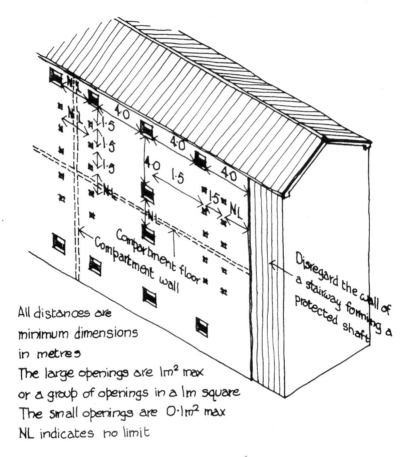

All distances are
minimum dimensions
in metres
The large openings are 1m² max
or a group of openings in a 1m square
The small openings are 0·1m² max
NL indicates no limit

Fig. 4.15 Unprotected areas which may be disregarded for separation distance purposes

4.12.4 Wall linings

The material with which an external wall is lined is the surface on which any fire will impinge and therefore its choice will significantly affect the growth and spread of a fire. This is particularly important in circulation spaces where any rapid spread is most likely to prevent the escape of occupants and it is the internal linings that are the main means by which it could occur.

The two properties on which guidance is given in Approved Document B are the rate of flame spread and the rate at which the material gives off heat when burning. There are two others that should be examined: the generation of smoke and the generation of fumes, either of which could hamper the evacuation of the building. Help on this should be sought from the manufacturer.

The selection of materials should be based on test results obtained through the methods given in BS 476: Part 6: *Method of test of fire propagation for products* and Part 7: *Method of classification of surface spread of flame of products*. The former produces a propagation index and the latter gives a classification between 1 and 4. Approved Document B also refers to a Class O. This is not defined in the Standard but is a material with a spread of flame classification of 1, a propagation index of not more than 12 and a sub-index of not more than 6. Class O is further explained in 6.4.

The Standard required to satisfy the requirements of the regulations is:

Location of lining	*Class*
Room of not more than 4 m² in a residential building	3
Room of not more than 30 m² in a non-residential building	3
Other rooms	1
Circulation spaces	O

Parts of the wall lining may have a lower classification than that given above (but not lower than Class 3) provided that the total area does not exceed either 60 m² in a non-residential building, 20 m² in a residential building, or one half of the area of the room, which ever is the less.

Wall lining materials are classified as follows:

Class O: Cement or gypsum-based plasters.
Plasterboards, unpainted, painted or with a PVC facing not more than 0.5 mm thick fixed directly to the wall or with an air gap.
Plasterboards as above with a fibrous or cellular insulating material.
Ceramic tiling.
Mineral fibre tiles or sheets with cement or resin binding.
Aluminium-faced fibre insulating boards.
Flame-retardant decorative laminates on a calcium silicate board.
Thick polycarbonate sheet.
Phenolic sheet.
Unplasticized polyvinyl chloride materials.

Class 1: Any of the materials listed as Class O.
Phenolic or melamine laminates on calcium silicate boards.
Flame-retardant decorative laminates on combustible boards.

Class 3: Timber or plywood boards with a density of not less than 400 kg/m³, unpainted or painted.
Wood particle boards or hardboard either treated or painted.

Chapter 5

GROUND FLOORS

5.1   The Building Regulations applicable

The Building Regulations relating to the construction of ground floors in buildings used for public assembly or residential purposes are:

A1   Loading
C1   Preparation of the site
C4   Resistance to weather and ground moisture
L1   Conservation of fuel and power

A1 requires that the building, in this case the ground floor, shall be constructed so that the imposed loads are sustained safely; C1 is concerned that the ground to be covered by the building shall be free from vegetable matter; C4 states that the floors of the building shall resist the passage of moisture to the inside of the building, and L1 requires that reasonable provision shall be made for the conservation of fuel and power.

5.2   Ground-supported floors

Ground-supported concrete floors comprise four elements: a hardcore base, a concrete bed, a damp-proof membrane and, in many cases, thermal insulation (see Fig. 5.1). Most floors are finished with a screed which is the finished surface or which is left ready to receive a final flooring, but this screed is not the subject of the Regulations except where the damp-proof membrane is laid on top of the concrete bed.

5.2.1   Structure

Turf, other vegetable matter and top soil should be removed from the ground to be covered by the building at least to a depth sufficient to prevent later growth as required by Regulation C1. This removal is usually taken to mean stripping away the surface down to the sub-soil level.

The excavated level can then be made up with hardcore. Clean broken brick or similar inert material, free from materials including water soluble sulphates in quantities which could damage the concrete bed are suitable materials. The maximum size of pieces of hardcore should be such as will pass through a 75 mm ring. No specific minimum thickness is given for the hardcore but the usual practice is to lay not less than a 100 mm depth and it should be well consolidated.

The bed of concrete to be laid on the hardcore should be at least 100 mm thick. It may well need to be thicker than this if the structural loadings are such as to require it. The loadings are given in BS 6399: *Loadings for buildings*.

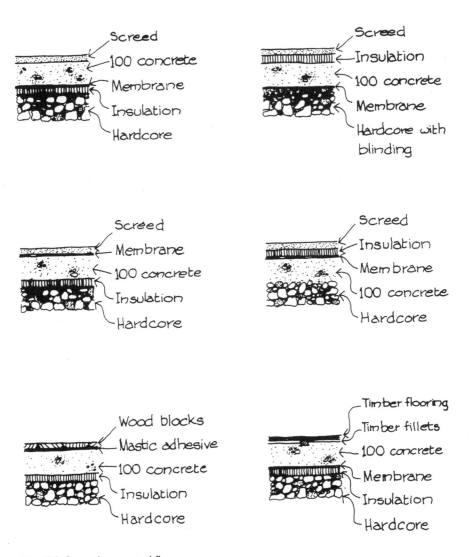

Fig. 5.1 Ground-supported floors

Part 1: *Code of practice for dead and imposed loads*. Heavily loaded floors may also need to be reinforced to distribute any point loads to be sustained. Typical floor loads to be allowed are shown in Table 5.1.

The concrete should be composed of:

50 kg of cement to
not more than 0.11 m³ of sand to
0.16 m³ of coarse aggregate.

Alternatively, it can be to BS 5328 mix designation ST2.

If there is embedded steel in the floor, the concrete should be composed of:

50 kg of cement to
not more than 0.08 m³ of sand to
0.13 m³ of coarse aggregate.

Alternatively, the concrete can be in accordance with BS 5328 mix ST4.

5.2.2 Damp-proofing

A damp-proof membrane must be provided within the floor construction unless it is a floor to which Regulation C4 does not apply, these are:

- A floor in a building used wholly for the storage of goods or the accommodation of plant or machinery, provided that any persons employed in the building are only there to store, service or remove the goods or plant.
- A floor in a building of such a purpose that the provision of a damp-proof membrane would not serve to increase protection to the health or safety of anyone habitually employed there.

The d.p.m. can be placed either under or on the concrete bed but it must be continuous with the damp-proof courses in the walls.

If it is laid below the slab it should be not less than 250 μm (1000 gauge) polythene sheet but, more often than not, a 300 μm (1200) gauge sheet is used. All joints between sheets should be sealed and the hardcore below must be finished with a bed of material which will not damage the polythene, i.e. blinded with sand.

To comply with Approved Document C and CP 102 (referred to below), a sandwich membrane laid on the concrete bed should be:

- 250 μm (or 300 μm) polythene, or
- three coats of a cold applied bitumen solution, coal tar/pitch or bitumen/rubber emulsion, or
- bitumen sheet damp-proof course to BS 743, or
- mastic asphalt to BS 1097 or BS 1418.

In all cases the membrane should be protected by a screed or other floor

Table 5.1 Floor loadings

Intended use of floor		Intensity of load	
		Distributed (kN/m²)	Concentrated (kN)
Assembly areas	Without fixed seating (dance halls, gymnasia, grandstands)	5	3.6
	With fixed seating*	4	–
Bars		5	–
Bathrooms, toilets, etc.		2	–
Bedrooms	Hospital bedrooms and wards	2	1.8
	Dormitories	1.5	1.8
Billiard rooms		2	2.7
Chapels and churches		3	2.7
Circulation areas (stairs, landings, corridors, aisles, footpaths, footbridges, terraces, plazas, etc.)	Subject to crowd loading	4	4.5
	Not subject to crowd loading	3	2.7
	Subject to more than crowd loading	5	4.5
Dining rooms		2	2.7
Dressing rooms		2	1.8
Drill rooms and drill halls		5	9
Equipment areas	General	2	1.8
	X-ray rooms, operating theatres, utility rooms	2	4.5
Exhibition areas (art galleries, museums, etc.)		4	4.5
Kitchens and laundries		3	4.5
Laboratories		3	4.5
Lounges		2	2.7
Plant rooms (boiler rooms, motor rooms, fan rooms)		7.5	4.5
Projection rooms		5	–
Reading rooms	With book storage (libraries)	4	4.5
	Without book storage	2.5	4.5
Stages		7.5	4.5
Storage	Dense mobile stacking	4.8/m stack height 9.6 minimum	7
	Book stack rooms	2.4/m stack height 6.5 minimum	7
	Stationery	4.0/m storage height	9
Vehicle areas (cars and vans maximum 2500 kg gross weight)	Driveways and ramps	5	9
	Workshops	5	9
	Parking areas	2.5	9

(*Source:* Based on Tables 5, 6 and 12 of BS 6399: Part 1: 1984)

* Fixed seating is seating where its removal and the use of the space for other purposes is improbable.

finish, and if either of the sheet materials is used it must be properly lapped and sealed. Polythene sheets require a lap of at least 150 mm and sealing with double-sided pressure sensitive tape; bitumen sheet damp-proof materials require a lap of at least 100 mm and sealing with bitumen lap cement.

The British Standard points out that certain floor finishes do not require a damp-proof membrane as either they themselves are damp-proof, such as coloured pitchmastic or mastic asphalt finishes, or are not adversely affected, such as terrazzo, concrete or clay tiles, cement/latex flooring or cement/bitumen flooring. The latter are not affected as they are capable of transmitting rising dampness to their upper surface where it can evaporate. This, of course, may not be compatible with the intended use of the floor.

The Standard also states that a polythene damp-proof membrane may not be adequate below floor finishes that are sensitive to moisture, either because of a danger that the adhesive might fail or dimensional or material changes may occur. Typical materials are magnesite, flexible PVC, sheet rubber or linoleum or cork, wood or chipboard flooring.

Alternatively, the membrane may be the adhesive bedding for a timber floor finish where it is of a type which would satisfy the requirement. If, instead of being laid in an adhesive bedding, the timber floor finish is fixed to fillets laid in the concrete, these must be treated with an effective preservative unless there is a damp-proof membrane below them (see Fig. 5.1). In this connection, the Approved Document refers to BS 1282: 1975 *Guide to the choice, use and application of wood preservatives*.

The required performance of a ground-supported floor can also be met by following the recommendations in Clause 11 of CP 102: 1973 *Protection of buildings against water from the ground*.

Where the circumstances are such that the floor will be subjected to water under pressure, the recommendation given in BS 8102 should be adopted. The most usual cause of water pressure under a floor is where it is below the water table level, as in the case of a basement. In these circumstances some form of tanking is recommended, as described in Chapter 3.

5.2.3 Thermal insulation

The standard thermal insulation requirement is a U value of $0.45\,\mathrm{W/m^2\,K}$. This insulation standard can be reduced (i.e. a higher rate of thermal transmittance) provided that there are compensating improvements in the thermal resistance of other elements in the building. A better insulation standard would allow a corresponding easing of the requirements in the rest of the building.

Overall, it must be possible to demonstrate that the total heat loss from the building will be no worse than it would be using the standard values. For this purpose either the Calculation Method or the Energy Use Method can be adopted. Details of these methods and their application are to be found in Chapter 9.

Table 5.2   *U* values of uninsulated ground floors

Ratio of perimeter to floor area	U value	Ratio of perimeter to floor area	U value
0.1	0.21	0.6	0.82
0.2	0.36	0.7	0.91
0.3	0.49	0.8	0.99
0.4	0.61	0.9	1.05
0.5	0.73	1.0	1.10

(*Source*: Based on Table C1 of Appendix C to Approved Document L (1995 edition))

The need for and thickness of insulation in a ground floor depends on its extent. The larger the floor the greater the area that is remote from the external temperatures and, hence, the less floor there is to lose heat. The rate of heat loss is a function of the ratio between the perimeter and the area. To find the *U* value of an uninsulated ground floor, divide the perimeter by the area, look up this ratio in Table 5.2 and read off the appropriate *U* value.

The insulation required to achieve a standard of 0.45 or 0.35 W/m² K can be found from Table 5.3 for both ground-supported and suspended ground floors. To use the table, first decide the *U* value to be achieved and then calculate the perimeter to floor area ratio. Opposite this figure in the appropriate section of the table can be found the minimum thickness needed by reading down from the appropriate heading and floor type. This may need to be rounded up to the nearest greater thickness available.

Table 5.3 has been based on three tables in Appendix A of Approved Document L (1995 edition) and reference should be made to these if a design *U* value of 0.25 W/m² K is required or if the insulation material to be used has a thermal conductivity other than those given in the table. Linear interpolation can be used to find the values for any material with a thermal conductivity between those quoted.

5.3   Suspended concrete ground floors

The use of suspended concrete ground floors has increased in recent years due, no doubt, to the combination of available economic systems and the inherent advantages of the constructional method.

5.3.1   Structure

The Approved Documents give no direct guidance on acceptable methods, materials, thicknesses or reinforcement for suspended concrete floors. How-

Table 5.3 Thickness of insulation required in ground floors

Insulation material	Thermal conductivity (W/m K)	Floor types	Required insulation thickness (mm) for a design U value (W/m² K) of: 0.35 — Length of the floor perimeter divided by the area of the floor									0.45 — Length of the floor perimeter divided by the area of the floor								
			1.0	0.9	0.8	0.7	0.6	0.5	0.4	0.3	0.2	1.0	0.9	0.8	0.7	0.6	0.5	0.4	0.3	<0.27
Phenolic foam board	0.02	Solid concrete	39	38	37	35	33	30	25	16	1	26	25	24	22	20	17	12	4	0
		Concrete beam	37	36	35	33	31	27	22	14	0	24	23	22	20	18	15	10	2	0
		Wood joist	57	55	53	51	47	42	34	22	1	37	35	33	31	27	22	15	4	0
Polyurethane board	0.025	Solid concrete	49	48	46	44	41	37	31	21	1	33	32	30	28	25	21	15	5	0
		Concrete beam	46	45	43	41	38	34	28	18	0	30	29	28	25	23	18	12	2	0
		Wood joist	67	66	63	60	56	50	41	26	1	44	42	40	37	33	27	18	5	0
Expanded polystyrene slab, glass fibre slab, mineral fibre slab	0.035	Solid concrete	68	67	65	62	58	52	43	29	2	46	44	42	39	35	30	21	6	0
		Concrete beam	64	63	61	58	54	48	39	25	0	42	41	39	36	32	26	17	3	0
		Wood joist	87	85	82	78	73	65	53	35	2	57	55	53	49	43	36	25	7	0
Glass fibre quilt	0.04	Solid concrete	78	76	74	70	66	59	49	33	2	53	51	48	45	40	34	24	7	0
		Concrete beam	74	72	69	66	61	55	45	29	0	48	46	44	41	36	29	19	3	0
		Wood joist	96	94	91	87	81	72	60	39	2	64	62	59	54	49	40	28	8	0

(Source: Based on Tables A9, A10 and A11 of Approved Document L (1995 edition))

Floor types
Solid Concrete   Ground supported in-situ concrete slab.
Concrete beam   Suspended concrete beam and block.
Wood joist   Suspended timber joists, at least 48 mm wide at 400 mm centres.

ever, the requirements of Regulation A1 will be met by following the recommendations of the British Standards listed below:

BS 6399: *Loading for buildings*: Part 1: *Code of practice for dead and imposed loads*

BS 8110: *Structural use of concrete*:
Part 1: 1985: *Code of practice for design and construction*.
Part 2: 1985: *Code of practice for special circumstances*.
Part 3: 1985: *Code of practice for singly reinforced beams, doubly reinforced beams and rectangular columns*.

5.3.2   Damp-proofing

There are two requirements to be met, the floor must adequately resist moisture reaching the upper surface and the reinforcement must be protected against the harmful effects of moisture.

By virtue of being suspended the floor should remain dry if there is an adequate resistance to rising damp in the supporting walls. If, however, the ground beneath the floor has been excavated to below the surrounding ground and will not be effectively drained, then a damp-proof membrane must be provided.

To protect the steel the concrete must afford at least 40 mm of cover if the floor has been cast in-situ and at least the thickness of cover required for 'moderate exposure' if the concrete is precast.

5.3.3   Gas ventilation

In those situations where there is a risk of gas accumulating under the floor which might lead to an explosion, a ventilated air space must be left below the concrete (or insulation if it is provided). The space must have a clear height of at least 150 mm and be ventilated by openings in two opposing external walls so arranged that there is a free flow of air to all parts. Approved Document C4 states that the openings should be large enough to give an actual clear opening of not less than the equivalent of 1500 mm^2 for each metre run of wall, which is equivalent to one 75 × 225 mm air brick with a free area of not less than 4500 mm^2 every thirteenth stretcher (see Fig. 5.2).

5.3.4   Thermal insulation

As explained above at 5.2.3, the standard for thermal insulation in a ground floor is a conductivity not exceeding 0.45 W/m^{2}K unless there are compensating better insulation standards elsewhere in the building. The need for thermal insulating materials in the construction is determined by the size of the floor: the larger the floor, the less the need to insulate it. Whether

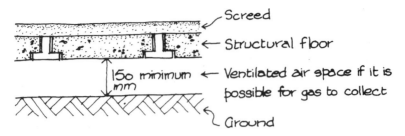

Fig. 5.2 Typical suspended concrete ground floor

insulation is needed and, if so, the thickness required can be found from Table 5.1, the use of which is explained in 5.2.3.

5.4 Suspended timber ground floors

Timber has been used for ground floors for a very long time but not always with success. The failure can nearly always be put down to a lack of observation of simple principles of construction and a lack of recognition of the fact that the material will decay in certain circumstances. The Regulation requirements represent what has been developed in the past as being good building practice. Timber is not commonly used in the construction of a ground floor in buildings used for public assembly but, provided that the loading conditions are observed, there is no structural reason why this material cannot be employed.

5.4.1 Structure

A timber ground floor will meet the performance requirements if:

- The ground is covered so as to resist moisture and prevent the plant growth below the floor.
- There is a ventilated air space between the ground covering and the timber structure.
- There is a damp-proof course between any material which can carry moisture from the ground to any timber.

The ground covering should consist of either:

- A 100 mm bed of concrete composed of 50 kg of cement to not more than 0.13 m³ of sand to 0.18 m³ of coarse aggregate (alternatively, mix ST1 of BS 5328 can be used if there is no embedded steel).
- A 50 mm bed of concrete composed as described above, laid on a sheet of polythene of minimum 1200 gauge; the polythene must have the joints sealed and be laid on a bed of material which will prevent damage to the sheet.

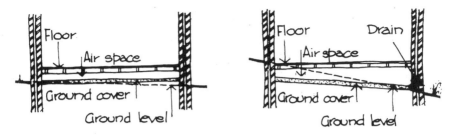

Fig. 5.3 Prevention of water collecting under suspended timber ground floors

To prevent water collecting below the floor and on top of the ground covering it should be laid so that its upper surface is above the level of the adjoining ground or, if this is not feasible, laid to falls with a drainage outlet through the wall above the lowest level of the ground (see Fig. 5.3).

The ventilated air space should measure at least 75 mm from the ground covering to the underside of any wall plates and at least 150 mm to the underside of the floor joists (or insulation if it is provided and lower than the joists). The ventilation is to be provided by openings in two opposite walls so that there will be a free flow of air to all parts. The openings should be large enough to give a free area of not less than the equivalent of 1500 mm^2 for every metre run of wall, or, in other words, a 75 × 225 mm air brick (with a free area of 4500 mm^2 or more) every thirteenth stretcher.

Any pipes required to duct the ventilating air to the underfloor space must have a minimum diameter of 100 mm.

The damp-proof course should consist of an impervious sheet material, such as engineering bricks in cement mortar, slates in cement mortar or any other material which will prevent the passage of moisture. It is usual to lay 50 or 75 × 100 m timber wall plates on the brickwork provided to support the floor joists and, if so, the damp-proof coursing should be placed immediately below these plates (see Fig. 5.4). Note that the joists should bear at least 35 mm onto the plates.

The sleeper walls, provided to support the wall plates should be honey-combed so that they do not prevent the free flow of ventilating air required and can be built off the ground covering described above, usually at approximately 1200 to 1800 mm centres.

The floor joists are fixed at 400 mm centres to coincide with the edges of 1200 mm flooring panels or, if floor boards are used, the spacing can be increased to 450 or 600 mm. The thickness of the flooring, size of the joists, their spacing and their span between sleeper walls depend on the load to be carried. This can be calculated from the details given in BS 5268: *Structural use of timber.* Appendix A of Approved Document A1/2 (1995 edition) gives tables of sizes for floor joists. These tables relate just to domestic work and should only be used with caution for floors in other buildings.

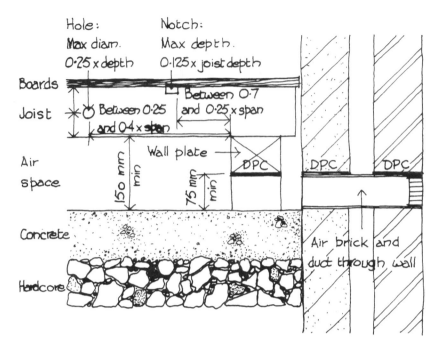

Fig. 5.4 Suspended timber ground floor

5.4.2 Notches and holes

Cutting away the joists for services is permitted, but only within the following limits:

- Notches should not be deeper than 0.125 times the depth of the joist and their position at a distance from the support of between 0.07 and 0.25 times the span of the joist.
- Holes should not be greater than 0.25 times the depth of the joist in diameter, drilled at mid-depth and not closer together than three times their diameter. They should be located at a distance from the joist of between 0.25 and 0.4 times the span of the joist (see Fig. 5.4).

5.4.3 Flooring

Appendix A refers only to softwood tongued and grooved floorboards as a finish to be laid over the joists in domestic work. The selection of the flooring to be used in a building intended for public assembly or residential use must be made on the basis of the loads to be carried, any high degree of wear due, say, to wheeled traffic, any spillages that might occur and any other uses that might be imposed on it.

5.4.4 Thermal insulation

The requirements for the insulation of suspended timber ground floors is exactly the same as that for suspended concrete floors given in 5.3.4 above.

Chapter 6

SUSPENDED UPPER FLOORS

6.1 The Building Regulations applicable

The Building Regulations relating to suspended upper floors in all buildings intended for public assembly and residential use are:

A1   Loading
B2   Internal fire spread (linings)
B3   Internal fire spread (structure)
E2   Airborne sound (floors and stairs)
E3   Impact sound (floors and stairs)
L1   Conservation of fuel and power

A1 requires that the building, in this case the upper floors, shall be constructed so that the combined dead, imposed and wind loads are sustained safely; B2 states that the internal linings of a building, in this respect, the ceilings, shall resist the spread of flame over their surface and, if ignited, have a rate of heat release which is reasonable in the circumstances; B3 has two main requirements in relation to upper floors – firstly, that the building shall be designed and constructed so that, in the event of fire, its stability will be maintained for a reasonable period and, secondly, to inhibit the spread of fire within the building, it shall be subdivided with fire-resisting construction to an extent appropriate to the size and intended use of the building. The latter requirement, it is pointed out in Approved Document B3, does not apply to material alterations made to any prison provided under Section 33 of the Prisons Act 1952; E2 and E3 stipulate that any floor or stairs separating a dwelling from another part of the building that is not part of the dwelling shall resist the passage of airborne sound and impact sound respectively; and L1 requires reasonable provision to be made to conserve fuel and power, in this application, by insulating floors exposed to the outside or unheated spaces.

6.1.1 Compartmentation

The principle of dividing the building into compartments is frequently referred to in connection with the control of fire. Approved Document B states:

The spread of fire within a building can be restricted by subdividing it into separate compartments by walls and/or floors of fire resisting construction. The object is two-fold:

- to prevent rapid fire spread which could trap occupants of the building; and
- to reduce the chance of fires becoming large, because large fires are more dangerous, not only to occupants and fire service personnel but also to people in the vicinity of the building.

As well as the subdivision of the building into compartments, it is also necessary to construct a compartment floor between those parts of the building that are in different occupancy or where the occupation is mainly for a different purpose. This does not apply, however, where one of those different purposes is ancillary to the other. An example of two main differing uses would be a hotel that contains a conference centre or similar accommodation where the public could assemble. Appendix D to Approved Document B defines ancillary use as where the area occupied by one use is less than a fifth of the area of the building or compartment devoted to the main use.

6.1.2 Fire resistance

The fire-resistance standard to be achieved in a floor in a building used for public assembly or for residential purposes depends on the following:

- The amount of combustible material in the building. This is referred to as the 'fire load' density.
- The height of the top floor above ground level. This affects the ease of escape, the efficiency with which fire-fighting operations can be effected and the consequences of a large-scale collapse.
- The occupancy and how it affects the speed with which the evacuation of the building can be carried out.
- Whether there are basements. These can increase the heat build-up because of their lack of an external wall through which to vent heat and smoke and they may also complicate the fire-fighting operations.
- Whether the building is of single storey. In this case the hazards are considerably less as evacuation would probably have been completed before any danger or structural failure occurs.

The Approved Document points out that the use of the building may change over its lifetime and with it the fire load. Precise fire engineering based on one specific use may be misleading and any future changes should be considered.

There are three criteria that apply to the fire resistance of materials and constructions:

- Resistance to collapse. This is referred to as the loadbearing capacity and, obviously, only applies to loadbearing elements.

- Resistance to fire penetration. This is referred to as the integrity and applies to any element that forms a fire separation.
- Resistance to the transfer of excessive heat. This is referred to as the insulation and also applies to any fire-separating element. It is particularly important in those parts of the building that enclose escape routes because a very high radiation of heat could prevent a person getting past the fire, even if the flames and smoke have been contained by the structure.

The degree of fire resistance is defined as the period of time, in minutes, during which the material or the construction continues to perform its function satisfactorily. The minimum periods required depend on the use of the building, whether it is fitted with an automatic sprinkler system meeting the relevant recommendations of BS 5306: Part 2 and the height of the top storey above ground level. The Building Regulation requirements for floors are set out in Table 6.1.

The Regulations also refer to BS 5588: *Fire precautions in the design, construction and use of buildings*: Part 1: *Code of practice for residential buildings* and Part 6: *Code of practice for assembly buildings*. Generally, the requirements of the Standards are less onerous than those of the Building Regulations but they do require the fire resistance to be achieved without any additional protection of the ceiling below.

6.2 Suspended timber floors

Timber is a popular material for the construction of suspended floors of buildings where the spans are small enough for it to be the most economic material from the aspect of strength/weight ratio and where there are no requirements for sound insulation or fire resistance. In the construction of buildings used for public assembly and residential use the sound insulation requirements are not always critical but fire resistance and load-carrying capacity frequently are such as to make other forms of construction more viable.

If a timber floor is required, the recommendations of BS 5268: *Structural use of timber*: Part 2: *Code of Practice for permissible stress design, materials and workmanship* and, for the fire resisting aspects, Part II of the Building Research Establishment's Report: *Guidelines for the construction of fire resisting structural elements* (BRE 1988) should be followed.

Where the proposals involve the conversion of an existing building, helpful advice on increasing the fire resistance of existing timber floors is given in BRE Digest 208 (BRE 1988).

6.2.1 Structure

The size of the structural timbers used as joists in a suspended timber floor is determined by the strength of the timber, the load to be carried, the span of the floor and the spacing of the timbers.

Table 6.1   Minimum periods (minutes) of fire-resistance for floors

Building use		Minimum fire-resistance period (minutes) of any floor, including a compartment floor					
		Basement storey, where the floor is below ground by:		Ground or upper storey, where the floor is above ground by:			
		>10 m	<10 m	<5 m	<20 m	<30 m	>30 m
Residential	Institutional	90	60	30*‡	60	90	120†
	Any other residential use	90	60	30*	60	90	120†
Assembly and recreational buildings	Fitted with sprinklers	60	60	30*	60	60	120†
	Not fitted with sprinklers	90	60	60	60	90	NP
Storage and other non-residential	Fitted with sprinklers	90	60	30*	60	90	120†
	Not fitted with sprinklers	120	90	60	90	120	NP

(*Source*: Based on Table A2 of Appendix A to Approved Document B (1992 edition))

* Where a compartment floor separates parts of the building in differing occupancies, the times marked should be increased to 60 minutes.
† Periods of 120 minutes apply to the structural frame only, the period for other parts of the structure is reduced to 90 minutes.
‡ Multi-storey hospitals, designed in accordance with NHS Firecodes, should have a minimum of 60 minutes as standard.
NP = Not permitted

Notes
1. Any floor over a basement should meet the provisions for the ground and upper storeys if that period is higher.
2. Where sprinklers are fitted they should be throughout the storey and in compliance with the relevant recommendations of BS 5306: Part 2.

For the purposes of the Building Regulations, there are two strength classes relating to timber: SC3 and SC4. Strength Class SC3 generally covers softwoods such as imported redwood and whitewood, home-grown fir, larch, pine and spruce as well as those same timbers from Canada and the USA all to grade 'general selected' (GS or MGS) or M50. Strength Class SC4 requires the same timbers to be to a grading of special selected (SS or MSS). Details of the timber species, their origin, the appropriate grading rules and the grades are to be found in Table 1 of Section 1B of Approved Document A1/2 or in BS 5268: Part 2: 1991.

Reference is made in Section 4.4 to the new Eurocode 5 that, eventually, will be the basis for the calculation of timber structures. While this is not applicable at the present time, engineers are recommended to be familiar with the changes it will impose and they may choose to work to them now.

The Approved Document sets out tables of joist sizes for floors but these are intended to relate to domestic work only with a floor load of $1.5 \, kN/m^2$. This loading only applies to the floors of dormitories in the range of building uses covered in this book. Table 5.1 sets out a full list of recommended floor load allowances for assembly and residential buildings as given in BS 6399: Part 1.

Because of the wide range of floor loadings to be accommodated, the timbers for non-domestic buildings should be calculated using the recommendations of BS 5268: Part 2. This is an exercise best left to a qualified structural engineer but a number of relevant points and principles are set out in Chapter 4. Note should also be taken of the new Eurocode 5, to which reference is made in Section 4.4.

6.2.2   Flooring

The Approved Document refers only to softwood tongued and grooved floorboards as a finish to be laid over the joists in domestic work. In non-domestic buildings, the loads, as explained above, can be much higher than in a house and the degree of wear is often much greater. These must be taken into account when selecting the type and thickness of flooring to be used.

6.2.3   Fire resistance

The fire-resistance standards for a floor exposed to fire from below are shown in Table 6.1.

The Approved Documents do not give any guidance on ways of achieving these standards but refer to specifications given in Part II of the Building Research Establishment's Report: *Guidelines for the construction of fire resisting structural elements* (BRE 1988). The following notes are taken from this publication.

Table 6.2 shows various constructions and their appropriate fire-resistance period. The table is based on joists of a width of 37 mm; if this is increased it may be possible to reduce the ceiling specification but guidance on this should be sought from the manufacturer concerned. The actual joist size must be based on calculations but with the condition that, for fire-resistance purposes, the maximum spacing for 9 mm plasterboard should be 450 mm but for 12.5 mm plasterboard it can be 600 mm.

Table 6.2 Fire-resisting construction for suspended timber floors

Construction and materials	Minimum thickness of protection (mm) for given floor finish and fire resistance:					
	Floor finish of 15 mm minimum*			Floor finish of 21 mm minimum*		
	Fire-resistance period (minutes)			Fire-resistance period (minutes)		
37 mm joists (minimum) with a ceiling of:	30	60	120	30	60	120
Metal lathing and sanded gypsum plaster of:	15			15		
Metal lathing and lightweight aggregate gypsum plaster of:	13	13	25†	13	13	25†
One layer of plasterboard with taped and filled joints, total thickness:				12.7		
Two layers of plasterboard with joints staggered, joints in outer layer taped and filled, total thickness:	22	31‡		19	31‡	
One layer of 9.0 mm plasterboard and sanded gypsum plaster finish of:	15			13		
One layer of 9.0 mm plasterboard and lightweight aggregate gypsum plaster (Mix V) finish of:	13			13		
One layer of 12.5 mm plasterboard and gypsum plaster of:	5			5		
One layer of 12.5 mm plasterboard and lightweight gypsum plaster (Mix V) finish of:	10			10		

(*Source*: Based on Table 14 of the BRE Report: *Guidelines for the construction of fire-resisting structural elements*)

* The floor finishes referred to are: tongued and grooved boarding, plywood and wood chipboard.
† Metal lathing to be also independently fixed with wire supports from the joist sides.
‡ In conjunction with 50 mm (min) wide joists and plywood or chipboard flooring.

Ceiling boards must be fixed to every joist and heading joints fixed to noggins. In double layer constructions, each layer must be fixed independently. Fixings are to be with galvanized nails, of the sizes shown below, at 150 mm centres:

9.5 mm plasterboard	30 mm nails
12.5 mm plasterboard	40 mm nails
19 mm to 25 mm plasterboard	60 mm nails

Expanded metal lath should be fixed at 100 mm centres with 38 mm nails or 32 mm staples. End laps should be not less than 50 mm, side laps not less than 25 mm and both wired together at 150 mm intervals. The lath should be spaced away from the background to provide a 6 mm gap to ensure adequate mechanical bond for the plaster.

6.2.4 Sound insulation

To comply with the requirements of Regulations E2 and E3 the floors of any building separating a domestic occupancy from a non-domestic use must resist the passage of sound generated by either a sound source such as a radio (airborne sound) or an object hitting the structure (impact sound). As well as the Regulation requirements, it may be desirable, or necessary, to incorporate sound insulation in the structure by virtue of its specific use or the requirements of legislation related to that use.

The means of providing resistance to each type of sound source is different. Airborne sound is absorbed by the mass of the floor, impact sound is reduced at source by a resilient floor surface. Since timber floors are a light form of construction, insulation against the direct transmission of sound is more difficult than is the case with concrete floors; nonetheless it is possible by the addition of pugging to increase the mass of the construction, but this may create difficulties in the case of alterations works due to the increased mass overloading the floor structure.

As well as the direct transmission of sound it is necessary to prevent flanking transmission round the edges of the floor and transmission through air gaps, such as could occur where pipes penetrate the structure.

Approved Document E defines three timber floor constructions considered to be satisfactory. These are reproduced in Fig. 6.1.

The specification of Floor A is a floating platform comprising a floor base of 12 mm softwood boarding or wood particle board, a 25 mm layer of mineral fibre with a density of $80 \, \text{kg/m}^3$ to $100 \, \text{kg/m}^3$, a layer of 19 mm plasterboard and either 18 mm tongued and grooved floor boarding or flooring grade particle board, in either case all joints glued and the flooring spot bonded to the plasterboard. An alternative floor finish would be two layers of cement-bonded particle board, glued and screwed together with the joints staggered.

The joists are normal structural joists of the size required for the span and an appropriate imposed loading allowance, taking into account the extra mass of the floor construction.

The soffit of the floor is finished with two layers of plasterboard, giving a total thickness of 30 mm, fixed with staggered joints on which is laid a 100 mm quilt of unfaced rockwool of a density not less than $10 \, \text{kg/m}^3$.

The specification for Floor B is a floating floor as for Floor A, nailed to 45 mm $\times$ 45 mm battens laid on resilient strips on top of the joists. As for

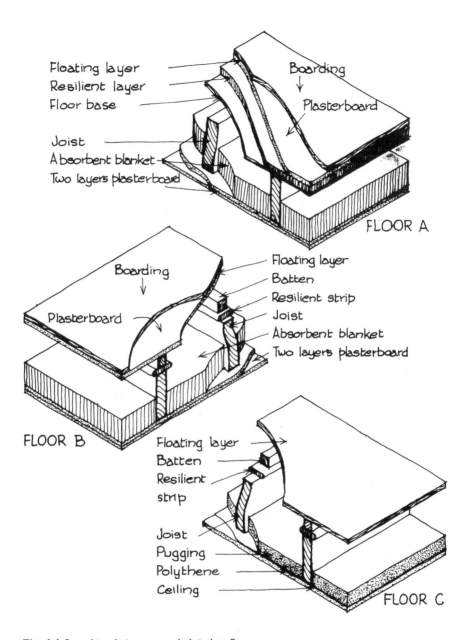

Floating layer
Resilient layer
Floor base
Joist
Absorbent blanket
Two layers plasterboard

Boarding
Plasterboard

FLOOR A

Boarding
Plasterboard

Floating layer
Batten
Resilient strip
Joist
Absorbent blanket
Two layers plasterboard

FLOOR B

Floating layer
Batten
Resilient strip
Joist
Pugging
Polythene
Ceiling

FLOOR C

Fig. 6.1 Sound insulating suspended timber floors

Floor A, the joists are normal structural timbers of a size appropriate for the loadings but not less than 45 mm wide. The soffit is finished with two layers of plasterboard and a 100 mm quilt all as specified for Floor A.

Floor C employs heavy pugging rather than, as in the case of Floors A and

B, an absorbent material. The floor finish is a normal 18 mm tongued and grooved boarding or flooring grade particle board, in both cases with the joints glued. The flooring is nailed to 45 mm × 45 mm battens laid in resilient strips on the joists. The joists must be not less than 45 mm thick and carefully sized to take account of the heavy nature of this floor.

The soffit of the floor is finished with a ceiling of either dense plaster, 19 mm thick, on expanded metal lath, or 6 mm plywood and two layers of plasterboard giving a total thickness of 25 mm.

Pugging is laid on the ceiling and consists of either a 75 mm bed of ash, a 50 mm bed of dry sand or a 60 mm bed of limestone chippings or whin aggregate, graded 2 mm to 10 mm.

Neither ash nor sand should be used in situations such as toilet floors where they may become wet and overload the structure.

Proprietary acoustic floor systems employ the same principle as Floor A, using a high density mineral wool bonded to high density tongued and grooved cement particle flooring laid on a deck of timber flooring, 22 mm thick on timber joists, finished on the soffit with a 31.5 mm thick plasterboard and plaster ceiling, which carries a 100 mm thickness of rockwool laid between the joists. This, it is claimed, exceeds the requirements of the Regulations in respect of sound insulation, provides a fire resistance of 60 minutes and offers a thermal resistance value.

To limit flanking transmission through the walls round the perimeter of the floor, all floating layers should be isolated from the face of the wall with a resilient strip (in the case of Floor A the resilient layer can be turned up the wall) and there should be a 3 mm gap below the skirting board. In addition the junction between the ceiling and the wall should be sealed with tape or caulked.

Any sound-insulating constructions differing from those given, which have already been built, tested as specified in Approved Document E, BS 2750: Part 4: 1980 and BS 2750: Part 7: 1980 and have been shown to be satisfactory can be repeated in other new work.

6.2.5   Thermal insulation

Any upper floor with its underside exposed to the external air must possess a U value of 0.45 W/m^{2}K (see Chapter 4 for an explanation of 'U value') and any upper floor over an unheated space, such as a garage or a store, must have a U value of 0.6 W/m^{2}K. The latter is referred to as a 'semi-exposed' floor.

The thickness of insulation required to achieve these standards can be calculated or can be found from Table 6.3. Table 6.3 shows an additional U value of 0.35 W/m^{2}K to demonstrate the effect of a higher standard of resistance should this be necessary in the design to compensate for lower standards in other elements in the building.

Table 6.3 Thickness of insulation (mm) required in suspended upper floors

Insulation material	Thermal conductivity (W/m K)	Floor type	Insulation thickness (mm) for a design U value (W/m² K) of:		
			0.35	0.45	0.60
Phenolic foam board	0.020	Timber	56	38	21
		Concrete	50	37	24
Polyurethane board	0.025	Timber	71	48	27
		Concrete	62	46	30
Expanded polystyrene slab, glass fibre slab, mineral fibre slab	0.035	Timber	99	67	37
		Concrete	87	65	42
Glass fibre quilt	0.020	Timber	114	76	42
		Concrete	99	74	48

(*Source*: Based on Tables A12, A13 and A14 of Approved Document L)

Floor types specification
Timber   Suspended timber joists, 48 mm wide at 400 mm centres
                 19 mm timber flooring
                 13 mm plasterboard
Concrete   Suspended concrete floor of any type
                 50 mm screed finish

6.3   Suspended concrete floors

Concrete floors are frequently employed in buildings used for public assembly or residential purposes because the standards of fire resistance and sound insulation are far more easily achieved.

6.3.1   Structure

Since a concrete floor is either purpose designed for its location or a proprietary system using prefabricated units there are no specific guidelines in any Approved Documents as to their construction, but Approved Document A1/2 lists the following British Standards as being relevant:

BS 6399: *Loading for buildings*:
    Part 1: 1984 *Code of practice for dead and imposed loads.*
BS 8110: *Structural use of concrete.*

Part 1: 1985 *Code of practice for design and construction.*
Part 2: 1985 *Code of practice for special circumstances.*
Part 3: 1985 *Code of practice for singly reinforced beams, doubly reinforced beams and rectangular columns.*

The calculations for a concrete floor are a task for a qualified structural engineer but relevant points and principles involved in the formation of concrete structures are set out in Chapter 4.

6.3.2 Fire resistance

The standards of fire resistance required to be achieved in a concrete floor depend on the use of the building and the height of the floor above ground level. The precise requirements are shown in Table 6.1.

No specific guidance is given in any Approved Document as to how to achieve these standards in a concrete floor, but the Approved Document requires that it should conform to an appropriate specification in Part II of the Building Research Establishment's Report: *Guidelines for the construction of fire resisting structural elements* (BRE 1988). The following notes are taken from this publication.

The fire-resistance properties of a concrete floor depend on the thickness of the slab and the thickness of the cover to the reinforcement, i.e. the distance from the heated face to the steel. The need for a certain slab thickness is obvious, the need to provide a minimum amount of concrete around the reinforcement is to ensure that the heat does not cause it to spall off thereby reducing the grip of the steel in the slab and weakening the strength of the structure.

Table 6.4 shows the thicknesses to be provided for different periods of fire resistance of a concrete floor with a flat soffit.

The slab thicknesses given in the table relate to solid floors; if it is a hollow slab or beam and block construction an effective thickness must be used instead, calculated from the actual thickness and the proportion of material per unit width. In these cases the manufacturer should be able to provide the necessary information.

6.3.3 Sound insulation

As stated above, any floor dividing a dwelling occupancy from a non-domestic use must be insulated against airborne and impact sound to comply with Regulations E2 and E3. In addition, it may be necessary to consider the sound insulating properties of the floor because of the specific use of the building.

Being of a greater mass than a timber floor, a concrete floor possesses an inherent sound-insulating value because the sound energy is absorbed by this greater mass. It may still be necessary to make special provisions to improve

Table 6.4   Fire-resisting in-situ suspended concrete floors

Construction and materials		Minimum dimensions (mm) (excluding any finish) for a fire resistance (minutes) of:			
		30	60	90	120
Reinforced concrete, simply supported					
Dense concrete	Thickness	75	95	110	125
	Cover	15	20	25	35
Lightweight concrete	Thickness	70	90	105	115
	Cover	15	15	20	25
Reinforced concrete, continuous					
Dense concrete	Thickness	75	95	110	125
	Cover	15	20	20	25
Lightweight concrete	Thickness	70	90	105	115
	Cover	15	15	20	20

(*Source*: Based on Table 10 of BRE Report: *Guidelines for the construction of fire resisting structural elements*)

this inherent value to meet the requirements of the Regulations and Fig. 6.2 shows two floor constructions considered to be satisfactory.

Floor A is composed of a structural slab with a resilient floor finish applied to it. The structural slab, including any screeding, permanent shuttering or ceiling finish is to have a mass of not less than $365 \, \text{kg/m}^2$ and the resilient finish is to have an uncompressed thickness of not less than 4.5 mm. The disadvantage of this method is its reliance on the floor finish to absorb impact sound. When new it will be satisfactory but after some time it will wear thin, thereby reducing its effect, or it may be replaced by another, non-resilient finish that will deprive the floor of its resistance to impact sound altogether.

Floor B does not have this disadvantage because the floating layer is not the floor finish and, therefore, any finish may be laid and, subsequently, re-laid without affecting the sound insulation performance. The structure of the floor is as for Floor A but its total mass need only be $300 \, \text{kg/m}^2$.

The floating layer can be either tongued and grooved timber or wood-based boarding fixed to $45 \times 45 \, \text{mm}$ battens or a cement and sand screed 55 mm thick with wire mesh reinforcement.

Either type of floating layer can be laid onto a resilient layer of mineral fibre of a density of $36 \, \text{kg/m}^3$, 25 mm thick. If the battens of the timber floating layer have an integral closed cell resilient foam strip, the thickness of the mineral fibre can be reduced to 13 mm. The fibre should be paper faced

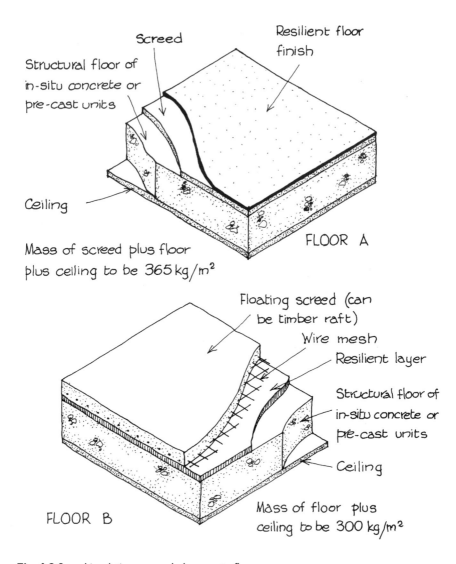

Fig. 6.2 Sound insulating suspended concrete floors

and laid paper face downwards under a timber finish and paper upwards under a screed.

Screeded floating layers can also be laid over pre-compressed expanded polystyrene board, 13 mm thick of the impact sound duty grade or extruded (closed cell) polyethylene foam of a density of 30 to 45 kg/m³, 5 mm thick.

The mineral fibre and the expanded polystyrene should be laid tightly butted and the joints of the polyethylene foam should be lapped. In all cases the insulant should be turned up at the edges to isolate the floating layer from the enclosing walls and the skirting board left with a small gap between its bottom edge and the floor.

Flanking transmission with both types of floor is dealt with in a similar way.

Where the floor adjoins an external or cavity separating wall, the mass of the wall leaf next to the floor should be at least $120 \, kg/m^2$ unless it is an external wall having openings of at least 20 per cent of its area in each room, in which case there is no minimum requirement. The floor base (excluding any screed) should pass through the leaf, whether it is supported by the leaf or not, but it should not bridge the cavity. If the floor base is constructed of concrete beams with infilling blocks or concrete planks, the first joint should be not less than 300 mm from the faces of the walls parallel to the structural elements.

Where the floor meets an internal partition wall with a mass of less than $375 \, kg/m^2$, the floor base, excluding any screeds, should pass through the wall. If the wall density is more than $375 \, kg/m^2$, it can pass through the floor and be tied to it with the joint grouted or, alternatively, the floor can pass through the wall.

6.3.4 Thermal insulation

All exposed and semi-exposed floors must be insulated to provide a resistance to thermal transmission of, in the case of an exposed floor, $0.45 \, W/m^2 \, K$ and, in the case of a semi-exposed floor, $0.6 \, W/m^2 \, K$. The thicknesses of insulation required to achieve these values can be found from Table 6.3 under the appropriate column for the insulation material to be used and the heading 'concrete'. The materials listed are those commonly available, but if other insulation materials are to be employed reference can be made to Table A13 of Approved Document L (1995 edition) that gives a larger range of thermal conductivities.

Additional screed thickness and floor finishes, such as wood blocks, can improve the thermal resistance of the construction and may lead to a reduction in the required thickness of insulation. Table 6.3 shows, an additional U value of $0.35 \, W/m^2 \, K$ to demonstrate the effect of a higher standard of resistance should this be necessary in the design to compensate for lower standards in other elements in the building.

6.4 Ceilings

The decorative aspects of ceilings are no concern of the Building Regulations but the fire aspects are. There are two points that are relevant in surface finishes to both ceilings and walls and they are the readiness with which the material will burn and the rate at which flame can spread across its surface.

The propensity of a material to propagate a fire is an important factor in its performance. The characteristics to be considered are its readiness to ignite, the rate of heat release and the time to the sudden eruption of fire known as 'flashover'. Fire propagation indices are found by reference to the test method specified in BS 476: Part 6. There are two results derived from this test, the Index of Performance (I) relates to the overall test performance and the sub-index (i_1) is found from the first 3 minutes of the test.

The classification of surface spread of flame is a measure of the rate at which a fire spreads across the face of the material under examination. The performance of a lining material is rated by reference to the test method specified in BS 476: Part 7 that classifies material according to a scale of 1, 2, 3 or 4 with Class 1 being the highest resistance.

The Approved Documents refer to another classification for fire perform-ance, Class O, and although this is not identified in any British Standard it is the highest product performance classification for a lining material such as a ceiling. It is achieved if the product is composed throughout of materials with limited combustibility or is a Class 1 material that has a fire propagation index (I) of not more than 12 and a sub-index of not more than 6.

Materials of limited combustibility are:

- Any non-combustible material such as totally inorganic products which, for linings, include metals and plaster and any other material classified under BS 476: Part 4 as non-combustible.
- Any material of a density of at least $300 \, \text{kg/m}^3$ which, when tested to BS 476: Part 11, does not flame and the rise in temperature on the furnace thermocouple is not more than 20 °C.
- Any material with a non-combustible core at least 8 mm thick and combust-ible facings to one or both sides not more than 0.5 mm thick. Where a spread of flame rating is specified, these facings must also meet the appropriate test requirements.

There is, in addition, a specification of limited combustibility that is only relevant to the insulation placed above a fire-protecting suspended ceiling, this is:

- Any material of a density of less than $300 \, \text{kg/m}^3$ that, when tested to BS 476: Part 11, does not flame for more than 10 seconds with a rise in temperature at the centre of the specimen of not more than 35 °C and a rise on the furnace thermocouple of not more than 25 °C.

Among the material used for ceilings, gypsum plaster, plasterboard (painted or not or with a PVC facing not more than 0.5 mm thick), woodwool cement slabs and mineral fibre tiles or sheets with cement or resin binding all meet both Class O and Class 1 rating.

Timber boarding or plywood with a density more than $400 \, \text{kg/m}^2$, painted or unpainted; wood particle board or hardboard either treated or painted and standard glass reinforced polyester products all are rated as Class 3. The timber products can be brought up to a Class 1 rating by an appropriate proprietary fire-resisting treatment.

Other materials, which may achieve the ratings shown, but need to be substantiated by test evidence, are:

- Class O: Aluminium-faced fibre insulating board, flame-retardant decorative laminates on a calcium silicate board, thick polycarbonate sheet and uPVC.
- Class 1: Phenolic or melamine laminates on a calcium silicate board and flame-retardant decorative laminates on a combustible board.

6.4.1 The required classification

The Building Regulation requirement for the classification of surface spread of flame in ceilings is the same as that for the surface of internal walls and is, for residential buildings, Class 3 in rooms of less than $4\,\mathrm{m}^2$; for non-residential buildings, Class 3 in rooms of less than $30\,\mathrm{m}^2$; Class O in circulation spaces; and Class 1 in any other rooms.

6.4.2 Suspended ceilings

Ceilings suspended below floors provide a ready route for the spread of smoke and flame through the cavity thus created. Approved Document gives several ways to limit this hazard:

- The cavity can be fitted with fire-resisting barriers to coincide with any other fire-resisting partitions within the building that divide the interior into compartments.
- If the compartment encloses a large area of the building, it may be necessary to install cavity barriers to limit the size of the ceiling void to 20 m in any direction unless the surface of the ceiling exposed to the void has a classification of less than Class O or Class 1, in which case the spacing of barriers is reduced to 10 m (8 m in an educational building).
- If the ceiling covers a single large open area greater than 20 m but not greater than 40 m the cavity barriers need only be installed along the lines of the walls enclosing the space.
- If the ceiling is over a single large space in excess of 40 m, it is still not necessary to install cavity barriers provided that:
  (a) the room and the cavity together are compartmented from the rest of the building
  (b) an automatic fire detection and alarm system is fitted throughout the building
  (c) the surface of the ceiling exposed in the cavity is Class O and it has non-combustible fixings
  (d) the recommendations of BS 5588: Part 9 are followed in respect of any use of the void as a plenum chamber for recirculating air

(e) the flame spread of any pipe insulation within the void is Class 1

(f) any electrical wiring in the void is laid in metal trays or run in metal conduit

(g) any other material in the void are of limited combustibility

(Details of the construction and fixing of cavity barriers are given in Chapter 8.)

- The ceiling itself may be fire resisting provided that its extent is not greater than 30 m. For this purpose, the ceiling should:

    (a) have a spread of flame rating for its upper surface, i.e. facing the cavity, of Class 1

    (b) have a spread of flame rating for its soffit of Class O

    (c) have a fire resistance of at least 30 minutes

    (d) be imperforate except for the passage of:
    - pipes (which should be properly fire stopped as described in Chapter 6)
    - fire-resisting ducts or ducts fitted with automatic fire shutters where they pass through the ceiling
    - cables or conduits containing one or more cables

    (e) extend throughout the building or compartment

    (f) not be demountable

6.4.3 Fire protecting suspended ceilings

A suspended ceiling may be taken to contribute to the overall fire resistance of the floor construction. If the floor and ceiling assembly has to meet a 60 minute fire resistance standard (see Table 6.1) the ceiling should be of a material of limited combustibility. Any access panels should be secured in position by screws or releasing devices and they should be shown to have been tested in the ceiling assembly in which they are incorporated. Any insulation in the ceiling void should be a material of limited combustibility.

6.4.4 Thermoplastic ceilings

Notwithstanding the requirements for fire resistance and spread of flame classification given in Section 6.4.1, the ceiling of a room may be constructed of thermoplastic materials that do not meet this classification.

Ceilings of this type may be panels in a suspended framework or may be in the form of a stretched skin – in either case of a grade of thermoplastic classified as TP(a) flexible. This grading applies to flexible products not more than 1 mm thick that comply with the Type C requirements of BS 5867: Part 2 when tested to BS 5438: Test 2. Each ceiling panel should not exceed 5 m^2 in area and should be supported on all sides and not be part of a fire-resisting ceiling.

6.4.5   Lighting diffusers

Lighting diffusers fitted as part of a ceiling obviously affect the fire-resisting qualities of that ceiling. By 'lighting diffusers', the Regulations mean any translucent or open-structured element which allows either natural or artificial light to pass into the room below. It does not refer to a lighting diffuser that is part of a luminaire suspended in the room below the ceiling.

If it is proposed to use such a lighting diffuser in a fire-protecting or fire-resisting ceiling, it must be shown to have been tested as part of that ceiling system and provide the appropriate fire protection.

Any other suspended ceiling, except one over a protected stairway, may incorporate an unlimited number of diffusers provided that the diffuser is of a thermoplastic classified as TP(a) rigid and the wall and soffit surfaces above the suspended ceiling comply with the general spread of flame requirements for the room as given in Section 6.4.1.

TP(a) rigid thermoplastics are:

- rigid solid PVC sheet
- solid polycarbonate sheet at least 3 mm thick
- multi-skinned rigid sheet of uPVC or polycarbonate with a Class 1 spread of flame rating
- any rigid thermoplastic product that, when tested to BS 2728: Method 508A, performs so that the test flame extinguishes before the first mark and does not flame or glow for more than 5 seconds after removal of the burner

If the diffuser is of a thermoplastic classified as TP(b) – that is, of rigid solid polycarbonate sheet less than 3 mm thick or multiple skinned polycarbonate that does not qualify as TP(a) – the maximum area of each diffuser, or the area of a group of small diffusers, is limited to $5 \, \text{m}^2$ with a space between the diffusers or group of diffusers of 3 m and a total area not exceeding 50 per cent of the floor area of the room (see Fig. 6.3). This type of diffuser can be placed over a circulation space with similar limitations except that the total area should not exceed 15 per cent of the floor area.

6.5   Pipes and ducts through floors

The necessity to provide protection where the passage of a pipe or duct through a floor creates an opening, only arises where the floor is required to possess fire-resisting or sound-insulating properties. In many situations, the floor must satisfy both criteria.

All pipes, except gas pipes, which penetrate a sound-insulating floor are to be enclosed in a duct above and below the floor. The construction of the duct has the following requirements:

- Material of a mass of at least $15 \, \text{kg/m}^2$.
- Lined internally with unfaced mineral wool, 25 mm thick or, alternatively, the pipe may be wrapped with this material.

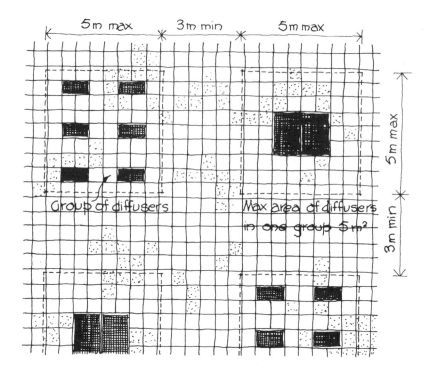

Type TP(b) lighting diffusers
set in a suspended ceiling

Fig. 6.3 Layout of lighting diffusers

- A 3 mm gap between the bottom of the duct and the floating floor surface, sealed with acrylic caulking or neoprene.
- Flexible fire stopping between the pipe and the edges of the hole in the floor so arranged as to prevent any rigid contact between either the pipe or the floor.

In the Gas Safety Regulations there are provisions for gas pipes to be in separate, ventilated ducts or they can remain unducted.

Pipes penetrating fire-resisting floors should be sealed with a proprietary fire seal, fire stopped, sleeved or encased in a duct. With a proprietary seal, the pipe can be of any diameter but if fire stopping is used the nominal internal diameter should not exceed:

160 mm   for a pipe in any non-combustible material
160/110 mm   for a stack pipe/branch pipe in lead, aluminium alloy, PVC or fibre cement and encased in a fire-resisting duct

40 mm	for a stack pipe or branch pipe as above, but not in a duct
40 mm	for a pipe in any material other than those above.

The fire stopping may be a proprietary material or cement mortar; gypsum-based plaster; cement- or gypsum-based vermiculite/perlite mixes; glass fibre, crushed rock, blast furnace slag or ceramic-based products (with or without resin binders), and intumescent mastics.

A pipe in lead, aluminium alloy, PVC or fibre cement may be run in a sleeve pipe through the floor that must project 1 m above and below the floor surface or soffit, be of non-combustible material and be in contact with the service pipe.

Where the pipe is to be encased in a fire-resisting duct all the internal surfaces, including the bounding wall faces, are to have a spread of flame classification of Class O. The casing itself should:

- have a fire resistance of not less than 30 minutes, including any access panels
- not be of sheet metal
- not have any access panels opening into a circulation space or bedroom
- be imperforate except for an opening for a pipe or access panel
- have the openings for pipes as small as possible and with fire stopping around the pipe

If the floor being penetrated is both fire resisting and sound insulating, a pipe casing must be provided which embraces both the above sets of requirements.

Chapter 7

ROOFS

7.1   The Building Regulations applicable

The Building Regulations relating to the construction of a roof of a building
intended as a place of public assembly or for residential use are:

A1   Loading
A4   Disproportionate collapse
B2   Internal fire spread (linings)
B3   Internal fire spread (structure)
B4   External fire spread
C4   Resistance to weather and ground moisture
F2   Condensation
H3   Rainwater drainage
L1   Conservation of fuel and power
Reg. 7 Materials and workmanship

A1 requires that the building, in this case the roof, shall be constructed so that
the imposed loads are sustained safely; A4 only applies to those parts of a
public building which have a roof with a span in excess of 9 m and is intended
to make sure that should any part of a roof fail, the building will not suffer a
collapse to an extent disproportionate to that failure; B2 is particularly
concerned with the ability of roof lights, plastic or otherwise, to resist the
spread of flame over their surface; B3 states that the building shall be
designed and constructed so that the unseen spread of fire and smoke within
concealed spaces in its structure and fabric is inhibited; B4 requires that the
roof of a building shall resist the spread of fire over the roof and from one
building to another; C4 requires that the roof shall resist the passage of
moisture to the inside of the building; F2 lays down that adequate provision
shall be made to prevent condensation in a roof or roof void above a
suspended ceiling; H3 is about the adequacy of systems carrying the rainwater
from a roof; L1 states that adequate provision shall be made for the
conservation of fuel and power; and Regulation 7 is concerned with the
standard of materials and workmanship employed.

While Regulation H3 relates to roofs, it also is a subject associated with drainage generally and is therefore dealt with in Chapter 14.

Regulation B3 has a significant historical significance since it was the rapidity with which fire spread through the roofs of houses in Pudding Lane that caused the initial ferocity of the Great Fire of London and gave rise to the first legislation on Building Control as we know it now.

In addition to the normal roofs provided to buildings, some structures, particularly those used for sporting pursuits, make use of an air-supported system. The Building Regulations give no requirements for this form of building; however, such a structure would be classed as a building and, therefore, subject to the rules laid down. There is guidance available in BS 6661 and this is reviewed in Chapter 4 at Section 4.9.

7.2 Weather resistance

The primary purpose of a roof is to exclude the weather. Approved Document C4, Section 5, treats roof finishes as cladding and sets out two requirements. Firstly, that it will resist the penetration of rain and snow to the interior of the building and, secondly, that it will not be damaged by rain or snow nor carry the rain or snow to any part of the building that would be damaged by it.

It is considered that any cladding with overlapping dry joints, i.e. slating and tiling, will meet the performance requirements if it is impervious or weather resisting and backed by a material that will direct any rain or snow that has entered the cladding back towards the outside face. This backing material is, of course, the sarking felt now laid below slating and tiling as standard practice.

Reference is also made in Document C4 to materials which deteriorate rapidly and which should only be used if certain conditions are met. The conditions attached to the use of such materials would be found in the relevant British Standards and Agrément Certificates of approval set out in the Approved Document supporting Regulation 7, reviewed in Chapter 1.

Where sheet roofing materials are to be used, the requirements can be met by following the recommendations of the relevant British Standard below:

BS CP 143: *Code of practice for sheet roof and wall coverings.*
  Part 1 deals with corrugated and troughed aluminium roofing.
  Part 5 deals with zinc roofing.
  Part 10 deals with galvanized corrugated roofing.
  Part 12 deals with copper roofing.
  Part 15 deals with sheet aluminium roofing.
  Part 16 deals with semi-rigid asbestos bitumen sheet roofing.
BS 6915: *Specification for design and construction of fully supported lead sheet roof and wall coverings.*

7.3   Structural stability

The load a roof is expected to carry due to snow lying on it varies, naturally, according to the area of the country in which it is located and the altitude of the site.

The map in Fig. 7.1 shows those parts of the country and altitudes where the imposed load can be taken as $0.75\,\text{kN/m}^2$ and those where it must be increased to $1.00\,\text{kN/m}^2$.

Appendix A to Approved Document A (1992 edition), gives tables of timber sizes for flat and pitched roofs. These are intended solely for roof members in single family houses. Even though the roof loadings and conditions are likely to

Fig. 7.1 Map of snow load allowances

be the same in, say, a small library building or a boarding house as they are in a private dwelling, the members of a timber roof intended for these purposes should be calculated in accordance with BS 5268: *Structural use of timber. Part 2: Code of practice for permissible stress design, materials and workmanship* or Part 3: *Code of practice for trussed rafter roofs.* Generally, the design of the roof should be left to a qualified structural engineer but the relevant details of the recommendations in Part 2 of the British Standard, given in Chapter 4 in relation to walls, also apply to roofs. The details given in Chapter 4 about the new European Discussion Document, DD ENV 1995–1: 1994 Eurocode 5: *Design of timber structures* should also be noted as they will, in time, become the standard basis for the calculation of roof frames as well as timber-framed walls.

The introduction to Part 3 of BS 5268 states that although trussed rafter roofs are intended primarily for dwellings they are also applicable to other buildings where the environment or service conditions are similar. It is also assumed that they are designed by a qualified trussed rafter designer. Two truss configurations are shown in the Standard: Fink and monopitch (see Fig. 7.2).

The possible spans depend on the species of timber used, its stress grade or

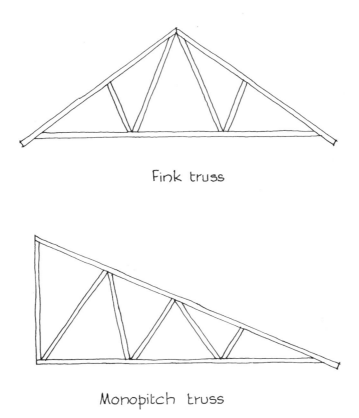

Fink truss

Monopitch truss

Fig. 7.2 Trussed rafter configurations

strength class (see Table 4.1), the pitch of the roof and the size of the members. The maximum span for a 35° pitch trussed rafter using 47 × 145 mm finished timbers is 12 m. This should not be exceeded, even where longer spans can be justified by calculation.

Trussed rafters should be fixed accurately at their design centres and should be braced to provide lateral restraint to ensure that the roof structure acts as a rigid component. The wall plates to which they are fixed should be wide enough to limit the compressive stress perpendicular to the grain. As a guide the width should be 0.008 × the span in millimetres with a minimum of 75 mm. In practice, plates with a width of 100 mm are commonly used and would be adequate for any span of roof up to the maximum recommended. Fixings between the truss and the plate are best effected by proprietary truss/plate fixings but, alternatively, two 4.5 × 100 mm nails can be used skew nailed into each side.

Hatches, openings for chimneys and the like are best designed to fit between the trussed rafter spacing but if this is not possible the spacing near the opening may be increased provided that it does not exceed either twice the design spacing or twice the design spacing minus the width of the opening.

7.3.1 Disproportionate collapse

In some building designs, the roof members are an integral part of the wall structure; a portal frame is an example of such a structure. In other buildings, the members of the roof are designed to afford lateral support to the walls. In either case, failure of a roof member could lead to an excessive failure of the walls. Section 6 of Approved Document A states that to meet the requirement of Regulation A4 in this respect, each structural member of the roof and its immediate supports must be considered as having been removed, one at a time, and the consequent effect on the structure analysed to ensure that the only associated failure is in any members supported by the one removed, although this may be accompanied by a substantial deformation of the building as shown in Fig. 7.3.

7.4 Thermal insulation and the control of condensation

As already explained in Section 4.10 dealing with the thermal insulation of external walls, the insulation of a building fabric is measured by its U value. Building Regulation L1 states that 'reasonable provision shall be made for the conservation of fuel and power'. Section 2 of Approved Document L1 states that this requirement will be met in relation to a normal pitched roof structure if the U value does not exceed $0.25\,\text{W}/\text{m}^2\,\text{K}$. If the roof is flat or is an insulated sloping roof with no loft space, the U value can be $0.45\,\text{W}/\text{m}^2\,\text{K}$.

The provision of sufficient insulation to achieve these standards of thermal

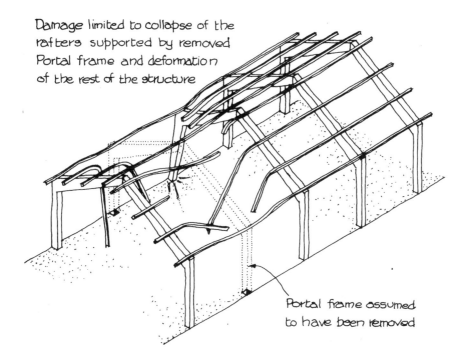

Damage limited to collapse of the rafters supported by removed Portal frame and deformation of the rest of the structure

Portal frame assumed to have been removed

Fig. 7.3 Acceptable extent of roof collapse

transmission is referred to in the Regulations as the Elemental Method. There are two other ways of showing that the proposed building will meet the requirements of the Regulation. These are the Calculation Method and the Energy Use Method. The first is a procedure whereby it can be shown that the rate of heat loss through the building fabric as a whole will be no greater than it would be through the fabric of a building of the same size and shape designed to comply with the Elemental Method standards. This allows a certain flexibility in the choice of materials and structures, permitting the use of a lower standard in certain elements of the building, where this is desirable because of other factors, providing that compensating higher standards are achieved elsewhere.

The Energy Use Method allows complete freedom of design using any valid energy conservation techniques, including taking account of useful solar and internal heat gains. As with the Calculation Method, the procedure for the Energy Use Method is to demonstrate, by calculation, that the annual energy consumption will be no greater than would be consumed in a similar building insulated to the Elemental Method standards.

Calculations for the Energy Use Method based on the CIBSE publication *Building Energy Code* 1981, *Part 2a* (worksheets 1a to 1e) are acceptable for this purpose.

The basic thicknesses of insulation required to achieve the necessary *U*

Table 7.1 Basic thickness of insulation in pitched and flat roofs

Insulation material	Thermal conductivity (W/m K)	Location of the insulation	Basic insulation thickness (mm) for a design U value (W/m² K)* of:		
			0.2	0.25	0.35
Phenolic foam board	0.02	A	167	114	69
		B	126	106	69
		C	97	77	54
Polyurethane board	0.025	A	209	142	86
		B	145	120	86
		C	122	97	68
Expanded polystyrene slab, glass fibre quilt, mineral fibre slab	0.035	A	293	199	120
		B	187	152	112
		C	170	135	95
Glass fibre quilt	0.04	A	335	227	137
		B	209	169	124
		C	194	154	109

(*Source*: Based on Tables A1, A2 and A3 of Appendix A to Approved Document L (1996 edition))

*See Table 7.2 for allowable reductions in the basic thickness arising from the insulation value of the roof components.

Notes
1. Location A: between the ceiling joists or rafters in a pitched roof or between the joists in a flat roof.
   Location B: between and over the ceiling joists in a pitched roof or between and over the joists in a flat roof.
   Location C: a continuous layer over a concrete roof or as a sandwich in a profile sheet roof. See Fig. 7.4 for illustrations of these locations.
2. The thicknesses given for A and B assume timber joists or rafters 48 mm wide at 600 mm centres.

values in pitched and flat roofs are shown in Table 7.1. Where the insulant is shown as located either 'between' or 'between and over' joists and rafters, they are assumed to be of timber in the proportion of 8 per cent, which is equivalent to 48 mm wide members at 600 mm centres. Table 7.2 shows allowable reductions that can be made in the basic thicknesses of insulation to take account of the insulation value of the various roof components listed. To arrive at a suitable insulation thickness, first obtain the basic thickness from

Table 7.2 Allowable reductions in basic insulation thickness for given roof components

Roof components	Allowable reductions (mm) in thickness of insulation materials in Table 7.1*			
	Phenolic foam board	Polyurethane board	Expanded polystyrene slab, glass fibre slab, mineral fibre slab	Glass fibre quilt
Pitched roofs				
Roof space	4	5	6	7
19 mm roof tiles	0	1	1	1
13 mm sarking board	2	2	3	4
12 mm calcium cilicate liner board	1	2	2	3
10 mm plasterboard	1	2	2	3
13 mm plasterboard	2	2	3	3
Flat roofs				
Roof space	3	4	6	6
19 mm timber deck	3	3	5	5
Concrete slab per 100 mm thickness of density of:				
600 kg/m³	11	13	18	21
800 kg/m³	9	11	15	17
1100 kg/m³	6	7	10	12
1300 kg/m³	5	6	8	9
1700 kg/m³	3	3	5	5
2100 kg/m³	2	2	3	3
50 mm screed	2	3	4	5
19 mm asphalt	1	1	1	2
3 layer built-up felt	1	1	1	2
10 mm plasterboard	1	2	2	3
13 mm plasterboard	2	2	3	3

(*Source*: Based on Table A4 of Appendix A to Approved Document L (1996 edition))

*See Table 7.1 for the basic insulation thickness.

Table 7.1 in relation to the type of insulant to be used and the U value to be achieved and then subtract each of the allowable reductions achieved by the roof components. The result will, probably, be a thickness that is not commercially obtainable, in which case, the nearest greater thickness must be used.

In roofs of the type using profiled roofing sheets fixed to a steel frame (such as one might employ in a sports hall) where the insulation is fixed between two layers of sheet roofing, the continuous layer location given in the table can be used. Section 7.5.1 gives details on the avoidance of concealed cavities as required by the Building Regulations. Figure 7.4 shows how this table is to be used to determine the thickness of insulation required in a steel-framed pitched roof, a timber-framed pitched roof and a flat concrete roof.

7.4.1 Condensation in roofs pitched at 15° or more

Approved Document F2 (1995 edition) gives guidance on minimizing condensation in roofs pitched at 15° or more where the ceiling is fixed to the soffit of the horizontal members of the frame and where the ceiling follows the line of the rafters to form a room in the roof. Further detailed guidance is given in the BRE Report BR 262: *Thermal insulation: avoiding the risks.*

Condensation is formed when warm, moist air meets a surface or point within a material where the temperature is low. This can often be seen as water running down a window pane. If it is allowed to happen in a roof, the resulting moisture can cause serious damage to the members of the structure.

There are two ways to reduce the dangers. Either the moisture-laden air must be prevented from reaching the colder parts by the use of vapour check membranes placed on the warm side of any insulation or else the damp air must be taken away, before it can cause trouble, by ventilating the spaces where it can collect. The former method, the Approved Document states, can reduce the amount of moisture reaching the roof void but it cannot be relied upon as an alternative to ventilation.

In a standard double-pitched roof, ventilation openings should be provided in both the eaves soffits, equivalent in area to a continuous gap of 10 mm wide.

The BRE Report recommends that there should also be a ventilation opening at the ridge when the span is greater than 10 m or the pitch is greater than 35° equivalent to a 5 mm continuous gap and that all eaves openings should be fitted with a 3 to 4 mm mesh or a proprietary unit to prevent the entry of insects.

If it is a monopitch roof where cross eaves ventilation is not possible, ventilation openings should be provided at as high a level as practicable, equivalent in area to a continuous gap of 5 mm.

Care should be taken to ensure that the air way from the eaves into the roof is not blocked by sagging sarking felt or thermal insulating materials, the BRE Report recommends the maintenance of an air path equivalent to a 25 mm

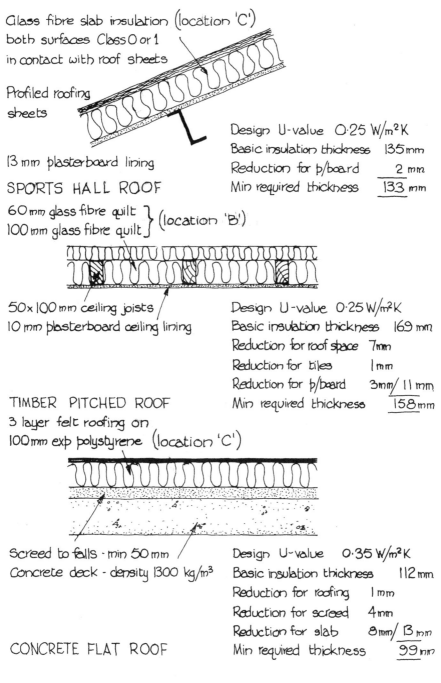

Glass fibre slab insulation (location 'C')
both surfaces Class 0 or 1
in contact with roof sheets

Profiled roofing
sheets

Design U-value   0·25 W/m²K
Basic insulation thickness   135 mm
Reduction for p/board   2 mm
Min required thickness   133 mm

13 mm plasterboard lining

SPORTS HALL ROOF

60 mm glass fibre quilt ⎫
100 mm glass fibre quilt ⎭ (location 'B')

50 x 100 mm ceiling joists
10 mm plasterboard ceiling lining

Design U-value   0·25 W/m²K
Basic insulation thickness   169 mm
Reduction for roof space   7 mm
Reduction for tiles   1 mm
Reduction for p/board   3 mm/ 11 mm
Min required thickness   158 mm

TIMBER PITCHED ROOF

3 layer felt roofing on
100 mm exp polystyrene (location 'C')

Screed to falls · min 50 mm
Concrete deck - density 1300 kg/m³

Design U-value   0·35 W/m²K
Basic insulation thickness   112 mm
Reduction for roofing   1 mm
Reduction for screed   4 mm
Reduction for slab   8 mm/ 13 mm
Min required thickness   99 mm

CONCRETE FLAT ROOF

Fig. 7.4 Thickness of roof insulation

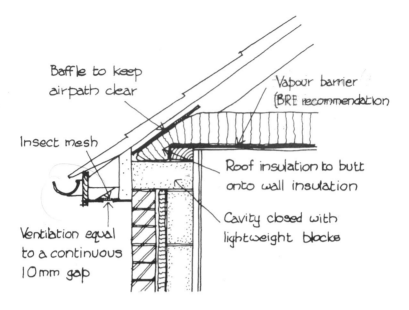

Baffle to keep
airpath clear

Vapour barrier
(BRE recommendation

Insect mesh

Roof insulation to butt
onto wall insulation

Cavity closed with
lightweight blocks

Ventilation equal
to a continuous
10mm gap

Fig. 7.5 Ventilation of pitched roofs

continuous gap at this point and both the Approved Document and the Report suggest the use of proprietary components to ensure this.

It is also recommended in the BRE Report that, to minimize the problem of condensation, all holes in the ceiling should be blocked so that there is no easy route for the moisture-laden air into the roof space. This involves carefully stopping up around all water pipes and electrical cables through the ceiling, particularly in areas of high humidity, and fitting any ceiling panels or hatches with a draughtseal that is efficiently compressed by bolts or catches.

There is also the possibility that a cold bridge can be formed at the eaves if the roof insulation does not meet the wall insulation or cavity closure and at the gables where there can be a small gap between the last ceiling joist and the wall.

Figure 7.5 shows details of a recommended construction for a timber-framed pitched roof.

7.4.2 Condensation in roofs pitched at 15° or more where the roof space is used

The task of ventilating a pitched roof is made more difficult where the ceiling and the insulation follow the line of the roof slope either fully or partly. To counter the restriction created by this construction, the eaves openings should be increased in size to an area equivalent to a continuous 25 mm gap and ventilation openings provided at the ridge equivalent to a continuous 5 mm gap. In addition, provision must be made to maintain a 50 mm wide airway up the slope between the insulation layer and the underside covering.

The BRE Report also recommends the use of a vapour control layer s 500 gauge (0.12 mm) polythene laid on the ceiling under the insulation a. breather membrane as a tiling underlayer. The report also points out ti. importance of not puncturing any lower walls within the roof space with cupboards.

7.4.3   Condensation in flat roofs

There are three forms of flat roof:

- Cold deck
- Warm deck (sandwich)
- Warm deck (inverted)

A cold deck roof has the insulation placed between the ceiling and the roof deck, a sandwiched warm deck roof has the insulation laid on a vapour barrier on the roof deck and is covered with the weather-proofing finish and an inverted warm deck roof has the insulation placed on top of the weather-proof layer. Figure 7.6 shows the three typical versions of these forms of construction.

The cold deck system is the one dealt with in Approved Document F2 that states that the Regulation requirement will be met by ventilation. The Document also states that it is not necessary to ventilate warm deck roofs or inverted roofs.

Satisfactory ventilation of a rectangular cold deck flat roof with a span of less than 10 m can be achieved by openings in two opposite sides equal to a continuous slot 25 mm wide running the full length of the eaves. If the span exceeds 10 m or the plan shape is not a simple rectangle, the ventilation openings should be increased to an area equal to 0.6 per cent of the roof area.

To allow a cross-ventilation to take place, an air path of at least 50 mm must be provided between the insulation and the roof deck.

The BRE Report also recommends that if the span exceeds 5 m the eaves ventilation should be increased to an area equivalent to a 30 mm continuous gap or 0.6 per cent of the roof area, whichever is the greater. It also states that the cross-ventilation airway should be increased to 60 mm when the span is between 5 and 10 m. A further recommendation of the report is that a vapour control layer of 500 gauge polythene or similar should be installed below the insulation and this should not be punctured by service pipes or cables. If it is required to run services at this level, a service void should be formed between the vapour control layer and the ceiling. The Approved Document does not require any vapour control layers (or vapour checks as the Document refers to

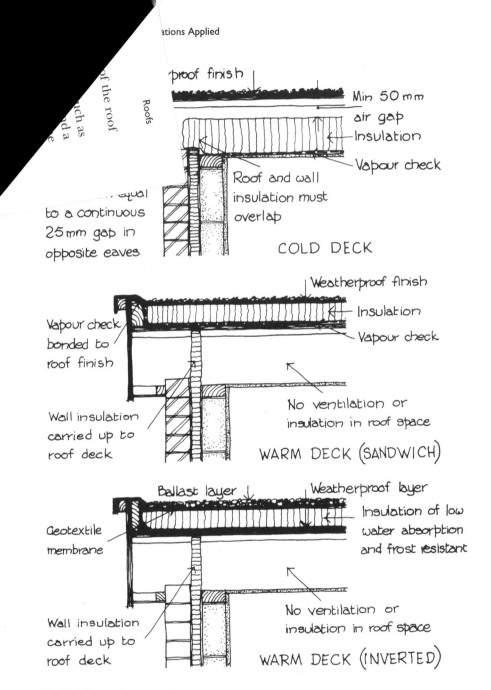

proof finish

Min 50 mm
air gap

Insulation

Vapour check

Roof and wall
insulation must
overlap

of the roof
ich as
d a
e

Roofs

equal
to a continuous
25 mm gap in
opposite eaves

COLD DECK

Weatherproof finish

Insulation

Vapour check

Vapour check
bonded to
roof finish

Wall insulation
carried up to
roof deck

No ventilation or
insulation in roof space

WARM DECK (SANDWICH)

Ballast layer

Weatherproof layer

Insulation of low
water absorption
and frost resistant

Geotextile
membrane

Wall insulation
carried up to
roof deck

No ventilation or
insulation in roof space

WARM DECK (INVERTED)

Fig. 7.6 Flat roof constructions

them) claiming that, while they can reduce the amount of moisture reaching
a void, they cannot be relied on as an alternative to ventilation.

Additional recommendations in the BRE Report are that insect mesh
should be provided to the eaves ventilators, any soil stack vent pipes should be

sealed to the ceiling and vapour control layer, the insulation must be carried over the wall plate to connect with the wall insulation and the cavity should be closed.

General recommendations in the Report are that flue outlets through an external wall should be at least 850 mm below any roof ventilation openings and one should only use an insulating walling block to close the cavity if there will be no significant movement between the leaves, otherwise it is suggested that a thin board of a suitable material such as calcium silicate or a roll of mineral fibre in a polythene cover should be used.

The Approved Document states that where the edges of the roof abut a wall or there are obstructions in the cross-ventilation airway (such as solid strutting or cavity barriers) or the movement of air outside the ventilation openings could be restricted, an alternative form of roof construction should be adopted. This alternative form is either a warm deck roof or an inverted roof. Details of these are given in the BRE Report BR 262 and the following notes are taken from this publication.

The warm deck, sandwich type of roof (see Fig. 7.6) is not so liable to condensation problems as the cold deck roof but placing the weather-proof membrane directly on the insulation can cause greater fluctuation in the surface temperature because on a sunny day the insulation prevents the roofing from losing its heat to the room and on a cold night it stops it gaining any warmth from below. For this reason a high tensile membrane to BS 747, Type 5 should be used if the finish is built-up felt roofing and the first layer should be partially bonded. An asphalt roof finish should be laid on sheathing felt and the insulant must be either selected for its resistance to the heat of the asphalt being laid or protected with a layer of bitumen impregnated fibreboard, corkboard or vermiculite board. As with the cold deck roof, provision must be made to connect the wall and roof insulation at the perimeter of the roof to prevent a cold bridge.

As the insulating material in an inverted warm deck roof is above the weather-proof membrane it is subject to a freezing/thawing cycle, wetting and the effects of ultraviolet light. Careful selection of the insulant is, therefore, essential, particularly with respect to water absorption and resistance to frost – the UV light problem is usually solved by the layer of ballast, which is required to prevent the insulation from blowing away, also preventing the light from reaching the insulation. The thickness of the insulation should be increased by 20 per cent to offset the cooling effect of rain running over the surface, particularly in winter. There is also a risk of localized cooling and condensation on the underside of lightweight roof decks if rain or snow can percolate down to the weather-proof layer. This is not usually a risk with concrete roofs because of their greater mass but with timber or metal-decked roofs it is recommended that the construction beneath the insulation possesses a thermal resistance of at least $0.15 \text{ m}^2 \text{K/W}$ and all insulation boards are tightly butted to each other and at projections or upstands. The geotextile membrane shown in Fig. 7.6 is required to contain the ballast and to prevent it from washing down between the insulating boards where it could, eventually, puncture the weather-proofing. This membrane must be turned up at all edges and abutments.

7.5   Fire resistance

Considerations of the dangers of fire with respect to a roof take three main forms: the spread of flame over the outside surface allowing the fire to carry to other buildings, fire within the roof spaces and cavities allowing it to spread to other rooms or buildings within the block and fire spreading across the ceiling and roof lights from one part of the building or room to another (particularly escape routes, see Chapter 15).

7.5.1   Roof coverings

The principal concerns are the effect on the coverings when exposed to fire from the outside and the distance from the boundary.

There are no limitations as to the distance from the boundary of any of the following roof systems:

- The traditional type of pitched roof constructed with timber framing, covered with sarking felt with or without any type of boarding and finished with slates, clay tiles, concrete tiles or fibre cement slates.
- A pitched or flat roof covered with mastic asphalt or any of the sheet metal roofings, including enamelled, alloy-coated or PVC-coated steel sheet, which are required to be fully supported. The support structure can be tongued and grooved or plain edged boarding on timber joists or rafters; woodwool slabs, compressed straw slabs, wood chipboard, fibre insulating board or 9.5 mm plywood on steel or timber joists or rafters; a concrete or clay pot roof slab or a non-combustible deck of steel, aluminium or fibre cement.
- A timber, steel or concrete pitched roof covered with either single- or double-skinned profiled roofing of galvanized steel, plain or PVC-coated aluminium or fibre-reinforced cement with or without underlays or insulation.
- A flat roof covered with bituminous felt on any type of roof deck provided that it is finished with bituminous-bedded stone chippings covering the whole of the surface to a depth of not less than 12.5 mm; bitumen-bedded non-combustible tiles; a sand and cement screed or a layer of tarmacadam.

If, however, the building has a pitched roof covered with built-up bituminous felt roofing there may be limits as to how close to the boundary it can be placed, depending on the type of felt to be used and the type of deck on which it is to be laid. Any type of felt upper or lower layers laid over a deck of compressed straw slabs or screeded woodwool slabs is acceptable at any

Table 7.3

Type of upper layer	Type of under layer	Minimum distance from the boundary
Type 1E	Type 1B: minimum mass 13 kg/10 m^2	6 m in a residential building under 1500 m^3, provided that the roof area is less than 3 m^2 and 1.5 m from a similar part
		12 m in a residential building over 1500 m^3
		12 m in an assembly building
		12 m in a storage building
Type 2E	Type 1B: minimum mass 13 kg/10 m^2	6 m in any building for public assembly or residential use
Type 2E	Type 2B	No limit
Type 3E	Type 3B or 3G	6 m in any building for public assembly or residential use

Note: The types of roofing felt referred to are described in BS 747.

distance from the boundary, but if the deck is of 6 mm plywood, 12.5 mm chipboard, 16 mm tongued and grooved timber boarding or 19 mm plain-edged timber boarding the distance to the boundary must be not less than that shown in Table 7.3.

Double-skinned profiled roofing systems can create a cavity between the sheets that could provide a ready route for the spread of fire or smoke. To overcome this hazard, Approved Document B3 states that the material of the sheeting should be of limited combustibility, both surfaces of the insulating layer should have a spread of flame classification of at least Class O or Class 1 and should make contact with both the outer and inner skins of the roofing. If any of these conditions are not satisfied, cavity barriers must be introduced as described in Chapter 8.

7.5.2 Roof structure

Generally, the structure that only has the function of supporting a roof is excluded from the fire-resisting standards imposed on elements of structure. The following exceptions apply:

Roof description	*Fire resistance*
A roof that is essential for the stability of an external wall that, itself, needs to have fire resistance	The same resistance as the wall

| A roof that also performs the function of a floor, such as a roof car park | The resistance shown in Table 4.9 |
| A roof that also affords a means of escape | 30 minutes |

In all cases the resistance is to an exposure to fire from the underside.

7.5.3 Ceilings

Section 6 of Approved Document B2 states that the ceiling of small rooms of not more than $30\,m^2$ in non-residential buildings can be of a material with a surface spread of flame classification of Class 3; other rooms should have ceilings that possess Class 1 grading and circulation space ceilings should be to Class O.

Gypsum plaster, plasterboard (painted or not or with a PVC facing not more than 0.5 mm thick), woodwool cement slabs and mineral fibre tiles or sheets with cement or resin binding all meet both Class O and Class 1 rating.

Timber boarding or plywood with a density of more than $400\,kg/m^3$, painted or unpainted; wood particle board or hardboard either treated or painted and standard glass-reinforced polyester products all are rated as Class 3. The timber products can be brought up to a Class 1 rating by an appropriate proprietary fire-resisting treatment.

Other materials that may achieve the ratings shown but need to be proved by test evidence are:

Class O: Aluminium-faced fibre insulating board, flame-retardant decorative laminates on a calcium silicate board, thick polycarbonate sheet and uPVC.

Class 1: Phenolic or melamine laminates on a calcium silicate board and flame-retardant decorative laminates on a combustible board.

Further details of the requirements for ceilings, both directly attached ceiling linings and suspended ceiling systems in relation to suspended upper floors are given in Chapter 6 and also apply to ceilings to the soffits of roof structures.

Lighting diffusers fitted as part of a ceiling significantly affect its fire-resisting properties and are subject to specific Regulation requirements. These are also set out in detail in Chapter 6.

7.5.4 Ceiling hatches

An access panel or hatch in a ceiling would provide a ready route for the spread of fire unless precautions are taken to ensure that its resistance is equal to that of the ceiling in which it is situated.

Regulation B2 requires that a hatch in a fire-resisting ceiling should have a fire resistance equal to that laid down for doors in cavity barriers, i.e. at least 30 minutes.

In fire-protecting ceilings, the form of hatch must be shown to have been tested as part of the ceiling system and to have achieved the same standard as that required for the ceiling.

As well as the potential loss of fire resistance, hatches leading to roof spaces can also provide a leakage path for cold air from the roof space into the building and, thereby, significantly affect the heating demand. To limit this effect, Approved Document L1 states that hatches to unheated roof voids should be fitted with draught seals and a suitable bolt or catch which will compress the hatch door into the seal.

7.6   Rooflights

Rooflights represent both a potential heat loss and a fire hazard.

The heat loss is limited by the imposition of the requirement that the rooflight should have a U value of not more than $3.3\,W/m^2\,K$ and an area of not more than 20 per cent of the roof area. This is the standard U value. It can be varied, but if the resulting heat loss exceeds the Standard, a corresponding saving must be made elsewhere in the building fabric.

The fire hazard aspects can be satisfied by ensuring that there is sufficient distance separating the rooflights and between the rooflights and the boundary to prevent the spread of fire to an adjoining premises. The boundary distances required depend on the material of the rooflight and its upper and lower surface fire-resistance classifications.

There are no restrictions from the fire-resistance point of view if the rooflight is made of unwired glass at least 4 mm thick, but a single sheet of this glass would fall very short of the thermal insulation requirements. Approved Document B makes no reference to the fire resistance of wired glass but, since unwired glass is acceptable, it must be assumed that there are no fire restrictions placed on wired glass rooflights either.

Thermoplastic rooflights may be classified as TP(a) rigid or TP(b). There is also a classification of TP(a) flexible but this would not apply to rooflights. A thermoplastic material is any synthetic polymeric product with a softening point below 200 °C when tested to BS 2782: Part 1: Method 120A.

TP(a) rooflights are those made from:

- rigid solid PVC sheet
- solid (as distinct from double- or multiple-skinned) polycarbonate sheet or uPVC with a Class 1 rating when tested to BS 476: Part 7 (see Chapter 8 for an explanation of this Classification)
- any other rigid thermoplastic product that, when tested to BS 2782: Method 508A, performs so that the test flame extinguishes before the first mark and the duration of flaming or afterglow does not exceed 5 seconds following removal of the burner

TP(b) rooflights are those made from:

- rigid solid polycarbonate sheet less than 3 mm thick
- multiple-skinned polycarbonate sheet
- any other product of which a sample between 1.5 and 3 mm thickness does not burn at a rate faster than 50 mm/min when tested to BS 2782: Method 508A

Thermoplastic rooflights of any classification are not permitted over a protected stairway, but over any other space a thermoplastic rooflight with an external surface classification of either TP(a) or TP(b) is acceptable but should be sited not less than 6 m from the boundary.

The under surface of rooflights determines their maximum permitted area and spacing. If the classification of the lower surface is TP(a), no limit is applied, but if it is TP(b) the maximum area of the rooflight (or a group of rooflights within a 5 m square) is 5 m². These individual rooflights, or groups of rooflights, must also be spaced 3 m apart and their total area should not exceed 50 per cent of the floor area of the space below, unless that space is a circulation area, in which case the percentage is reduced to 15 per cent. This spacing is the same as that given for lighting diffusers and is illustrated in Fig. 6.3.

Chapter 8

INTERNAL WALLS

8.1   The Building Regulations applicable

The Building Regulations concerned with the standards of construction of internal walls in any buildings to be used for the purpose of assembly or as a residential premises are:

A1   Loading
B2   Internal fire spread (linings)
B3   Internal fire spread (structure)
E1   Airborne sound (walls)
L1   Conservation of fuel and power
N1   Glazing

A1 requires that the building, in this case any loadbearing internal walls, must be constructed so that the loads are sustained and transmitted safely to the ground; B2 states that the internal linings, such as the wall surfaces, must resist the spread of flame over their surface and, if ignited, have a rate of heat release which is reasonable in the circumstances; B3 has four requirements, (1) the building shall be constructed so that its stability will be maintained for a reasonable period in the event of a fire, (2) a wall common to two buildings shall resist the spread of fire between those buildings, (3) the building shall be subdivided by fire-resisting constructions, which in part will be the internal walls, to inhibit the spread of fire within the building, (4) the design and construction must inhibit the unseen spread of fire and smoke within concealed spaces; E1 states that a wall which separates a dwelling from another part of the building not used exclusively as part of that dwelling shall resist the transmission of airborne sound; L1 calls for the conservation of fuel and power by limiting heat loss through the fabric which, in the case of internal walls, relates to walls separating heated from unheated spaces; N1 requires that any glazing with which someone could come into contact must be safe and, if people are likely to collide with it, it must incorporate features that will make it apparent.

8.1.1   Other legislation

All buildings intended for public assembly must comply with the Building Regulations but the details of how this is to be achieved are set out in British

Standard 5588: Part 6, to which the Regulations refer. Where relevant, these details are quoted in this chapter. One of the most complex of the residential buildings subject to the Regulations is a hospital. In this case relevant details are to be found in the NHS Estate's Firecodes and Health Technical Memorandum, as well as in the Approved Documents. A similar document to the NHS Estate's Firecode is the Building Bulletin 7, produced by the Department of Education and Science with the title of *Fire and the design of educational buildings*. Where relevant, details from this publication are also included.

8.1.2 Types of internal wall

The Building Regulations recognize a number of different types of internal wall depending on their particular function and apply appropriate standards to each. The types of wall are:

Separating walls:	Walls that are common to two or more buildings.
Compartment walls:	Walls that divide a building into fire-resisting compartments to restrict the spread of a fire.
Loadbearing walls:	Any wall that supports the floors or roof of the building.
Buttressing walls:	Walls that are built at right angles to another wall, internal or external, and provide it with structural restraint.
Partition walls:	Walls that subdivide the space enclosed by the external walls of the building but do not support any load other than the self-weight of the partition.

8.2 Compartmentation

Compartmentation is the term given to the dividing of the interior of a building into enclosed spaces that are separated by fire-resisting floors and walls. The fire resistance of floors is dealt with in Chapter 6 and the form of fire-resisting compartments in a Nucleus type of Hospital design is described in Chapter 4.

The purpose of compartmentation is, firstly, to prevent the rapid spread of fire that could trap the occupants of the building and, secondly, to reduce the chance of the fire becoming large, on the basis that large fires are more difficult to deal with and are of greater danger than small ones.

The appropriate degree of division within the building depends on:

- The use of, and the fire load in, the building. This affects the potential for a fire to start, the severity to be anticipated and the ease with which evacuation can be effected.
- The height of the top storey above ground. This affects the ease of evacuation and the ability of the fire services to attend to the fire.

- The availability of a sprinkler system. This will affect the rate of growth of the fire and may even suppress it altogether.
- The building use. Schools and hospitals in particular require very careful consideration of the problems peculiar to their function.

8.2.1 The provision of compartmentation

The requirements for dividing a non-residential building into compartments depend on the occupancy of the building and the size of the enclosed space.

The maximum area of floor in any one storey of a building to be used for public assembly or recreation is $2000\,\text{m}^2$, unless a sprinkler system is fitted, in which case the maximum area can be doubled to $4000\,\text{m}^2$. If the total area of the building is greater than these limits it must be subdivided by compartment walls.

No maximum areas or volumes are given for residential buildings generally, except for hospitals where a limit of $2000\,\text{m}^2$ is set for multi-storey buildings and $3000\,\text{m}^2$ for single storey. The NHS Estates Firecode on Nucleus design hospitals points out that, because the main elements in this design (the cruciform compartments shown in Fig. 4.14) are based on four modules measuring $16.2 \times 16.2\,\text{m}$ the maximum area becomes $1500\,\text{m}^2$.

Building Bulletin 7: *Fire and the design of educational buildings*, states that in a single-storey school building either the area of the building or a compartment within the building should not exceed $800\,\text{m}^2$ or the volume should not exceed $8500\,\text{m}^3$. In multi-storey school buildings the area limit of $800\,\text{m}^2$ applies throughout to any storey or compartment.

In all buildings the following walls must be constructed as compartment walls with the fire resistance as shown in Table 8.1.

- Any wall needed to subdivide the storey where the floor area exceeds any of the maxima given above.
- Any wall common to two or more buildings.
- Any wall used to divide the building into separate parts that can be individually assessed for the purpose of determining the appropriate standard of fire resistance.
- Any wall enclosing and separating a cruciform-shaped clinical template in a Nucleus type of hospital.
- Any wall bounding a stairwell or lift shaft.

8.2.2 Design of compartment walls

Not only must the materials and construction of compartment walls ensure that they possess the degree of fire resistance indicated in Table 8.1 or Table 8.2, they must also be designed to form a complete barrier to fire between the compartments they serve.

In addition, any compartment wall that is common to two or more

Table 8.1   Minimum periods of fire resistance of loadbearing walls

Building use	Minimum fire-resistance period (minutes) for loadbearing walls					
	Basement storey, where the floor is below ground by:		Ground or upper storey, where the floor is above ground by:			
	>10 m	<10 m	<5 m	<20 m	<30 m	>30 m
Assembly and recreation						
Not fitted with sprinklers	90	60	60	60	90	NP
Fitted with sprinklers	60	60	30*	60	60	120†
Storage and other non-residential						
Not fitted with sprinklers	120	90	60	90	120	NP
Fitted with sprinklers	90	60	30*	60	90	120†
Residential (except a hospital or educational building)	90	60	30*	60	90	120†
Hospital						
Not fitted with sprinklers	90	60	30	60‡	90	120
Fitted with sprinklers	60	60	30	60†	60	90

	Number of storeys					
	Basement		Ground and upper floors			
	7 to 5	4 to 1	1	1 to 4	5 to 7	>7
Educational (maximum floor area 800 m²)	90	60	Nil§	30	60	90

(*Source*: Tables A1 and A2 of Appendix A to Approved Document B (1995 edition), NHS Estates Firecode. Health Technical Memorandum 81: 1996 and DES: *Fire and the design of educational buildings*)

NP = Not permitted

Notes
1. If the wall is a compartment wall separating buildings, the times marked * must be increased to 60 minutes.
2. If the wall is a compartment wall separating occupancies, the times should be 60 minutes or as shown in the table, whichever is the less.
3. In all cases, the times given for a compartment wall apply to its loadbearing capacity, integrity and insulation but, for a loadbearing wall the times apply to its loadbearing capacity only.
4. Where the wall does not form part of the structural frame, the times marked † can be reduced to 90 minutes.
5. In a hospital, the height of the upper floor at which the times marked ‡ apply is 12 m.
6. Where sprinklers are fitted, they should be throughout the storey and in compliance with the relevant recommendations of BS 5306: Part 2.
7. The time marked § must be 30 minutes where the 8500 m³ maximum volume rule applies or where the wall is a compartment wall, supports a gallery or encloses a protected corridor.
8. Fire-rated stores in schools (chemical, oil and PE equipment stores) must have 60 minute construction.

Table 8.2 Periods of fire resistance in specific walls

Building use	Location of wall	Fire-resistance period (minutes)	Comments
Hospital	Compartment wall	60	
	Sub-compartment wall	30	Sub-compartment walls separating high hazard areas such as Intensive Therapy Units should be 60 minutes
	Enclosing a hospital street	60	
	Between a normal dependency patient access area and: a boilerhouse, stores, laundries, kitchens, main electrical switchgear, refuse collection or incineration areas	60	Sprinklers must be installed in the hazard area. These areas must not be located adjacent to very high dependency patient areas. Medical gas should be stored in a separate building
	Between a normal dependency access area and: central staff changing rooms, sterilization and disinfection units, pathology laboratories, manufacturing pharmaceutical departments	60	These areas can be located adjacent to a very high dependency patient area but, if so, sprinklers must be fitted in the hazard area
	Subdivision within a laundry	30	
	Subdivision within a kitchen	30	Applies to walls enclosing the kitchen, dining room, servery, store and changing rooms
Places of Assembly	Loadbearing walls	30	Applies to loadbearing capacity, integrity and insulation
	Non-loadbearing walls	30	Applies to integrity and insulation only
	Walls enclosing: dressing rooms, kitchens, staff restaurants and canteens, projection rooms for non-flammable film, storage areas less than $450\,m^2$, repair or maintenance workshops, low-voltage equipment rooms	30	
	Walls enclosing any other ancillary accommodation	60	
	Proscenium wall	60	From the lowest level of the Stage Basement to the underside of the roof

(*Source*: Based on BS 5588: Part 6 and NHS Estates Firecode, Health Technical Memorandum 81: 1996)

buildings or that forms a separate part of the building must run the full height of the building in a continuous vertical plane so that the division is solely by a compartment wall, not by a floor.

A proscenium wall must run, as shown in Table 8.2, from the lowest level of the Stage Basement to the underside of the roof. However, all proscenium walls must have a large opening in them, where the stage is built. This opening is protected by a safety curtain and the necessary details of these are given in Chapter 9.

8.3   Structure

While the Building Regulations offer detailed guidance on the construction of internal walls in buildings for residential use, the requirements for the building types dealt with in this book rest firmly on the recommendations of British Standards.

Section 02 of Approved Document A1/2 states that the safety of a structure depends on the successful combination of design and construction. Loads used should allow for all possible dynamic, concentrated and peak loads that might occur and should be in accordance with BS 6399: Part 1 and Part 3 with respect to dead and imposed loads and CP 3: Chapter 5: Part 2 with respect to wind loads. These, in due course, will be replaced by the appropriate Eurocode (see Chapters 3 and 4). Besides loading, the other aspects that must receive detailed consideration are the properties of the materials, details of the construction, workmanship and safety factors and these must all be incorporated in a design analysis.

In most buildings used for assembly purposes and many used for residential purposes, the structure is a frame that may or may not rely on certain internal walls for stiffness. Generally, designs aim to avoid fixed internal divisions of a structural nature, thus making the accommodation as flexible as possible. The only regular exceptions to this are the enclosures to stair and lift wells and walls enclosing escape routes or sanitary accommodation.

Lift and stairwell enclosures need to be considered from a fire-resistance aspect (see 8.3) but, because they tend to be like a large hollow column extending the full height of the building, they may also serve a structural function. In this case they would be designed in with the rest of the structural system and in accordance with the recommendations of BS 6399: *Loading for buildings* and either BS 5628: *Code of practice for use of masonry* or BS 8110: *Structural use of concrete*. These are dealt with in detail in Chapter 4.

Escape route walls, like stairwell walls, need to be considered from a fire-resistance point of view but they do not usually perform any structural function.

The only structural consideration in partitions enclosing sanitary accommodation is the need to be strong enough to support the load of any appliances attached to them.

Any other internal partitions usually serve no other purpose but to subdivide space and, therefore, are not subject to any structural considerations.

8.4   Fire-resisting walls

Fire-resisting internal walls are required to separate the building into compartments to control the spread of fire. They are also required to enclose an escape route corridor, a protected space such as a stairwell, corridors serving any ancillary accommodation in an assembly building and dead-end corridors.

The fire resistance of a wall is measured in three ways: the length of time it can maintain its loadbearing capacity (resistance to collapse); the length of time it can resist the penetration of fire (integrity); and the length of time it can resist the transfer of excessive heat (insulation).

The periods of fire resistance, expressed in minutes, required in compartment walls and loadbearing walls that are not compartment walls are shown for buildings generally in Table 8.1 and in specific locations in Table 8.2. In addition, any other internal wall that encloses a protected stairway, a lift shaft, a service shaft, a protected lobby or corridor or that subdivides a corridor must possess a fire resistance of 30 minutes in respect of its loadbearing capacity, its integrity and its insulation. A restricted amount of glazing is permitted in these walls as described in Section 8.7.5.

The achievement of these fire-resisting standards is readily met by normal brick or block partitions, the resistances of which are as set out below. All values are given for unfinished walls since the application of plaster makes insufficient difference to have any effect upon the practical thickness that can be built (but see Section 8.3.4 on wall linings and spread of flame).

Loadbearing walls:
Half brick wall	120 min
100 mm lightweight block wall	120 min
100 mm aerated block wall	120 min

Non-loadbearing walls:
75 mm lightweight block wall	120 min
50 mm lightweight block wall	30 min
63 mm aerated block wall	120 min
50 mm aerated block wall	60 min

The fact that these thicknesses are adequate for the purposes of fire resistance does not mean that they are necessarily the right thickness to be built. Their structural stability and their thermal insulation may call for a greater thickness and must be considered.

Compartment walls in hospitals must be constructed of materials of limited combustibility as listed below:

147

- Products classified as non-combustible under BS 476: Part 4: 1970.
- Totally inorganic materials such as concrete, fired clay, ceramics, metals, plaster and masonry containing not more than 1 per cent of organic material.
- Any material of a density of at least $300 \, kg/m^3$ or more which does not flame or rise in temperature more than $20 \, °C$ when tested in accordance with BS 476: Part 11.
- Any material with a non-combustible core at least 8 mm thick having combustible facings on one or both sides, not more than 0.5 mm thick (plasterboard is an example of such a material).

8.4.1 Junctions of fire-resisting walls to other structures

For the fire-resisting wall or compartment wall to be effective and to offer a complete barrier to fire, there must be continuity at the junctions with the other fire-resisting elements of the building.

Any junction between fire-resisting walls or between a fire-resisting wall and an external wall must maintain the fire resistance of the walls.

Any junction between a fire-resisting wall and a roof needs to be fire stopped by carrying the wall up to the underside of the roof covering and filling the joint between the two with a resilient fire stopping that will maintain the continuity of the fire resistance. This fire stopping will prevent the fire spreading past the edge of the wall between the roof and the structure but there is also the possibility that it could spread by breaking through the roof adjacent to the wall. To restrict this, a zone of roof, 1.5 m wide on each side of the wall should have a covering of designation AA, AB or AC (see Section 7.5.1) on a deck of limited combustibility (see Fig. 8.1a). Alternatively, the roof covering and deck could be of a composite structure such as profiled steel cladding, with the requisite resistance.

This requirement is relaxed slightly in the case of a residential building other than one of an institutional nature and a building for public assembly, but in both cases they must not exceed a height of 15 m. The relaxation is that, while the roof covering must be designated AA, AB or AC, the deck of limited combustibility is not required and any timber tile battens may be carried over the wall provided that they are fully bedded in mortar for the full width of the wall and the wall is fire stopped up to the underside of the roofing. If the roof is boarded with combustible boards or wood wool slabs, these may also be carried over the wall, in which case the fire stopping is carried up to the underside of the boarding. A further relaxation is that roof support members may also pass through the wall but should be treated with a fire protection for a distance of 1.5 m from each side of the wall.

An alternative to all the above provisions, harking back to the oldest building by-laws of all, is given in Approved Document B3 that states that a compartment wall may be extended up through the roof to a height of at least 375 mm above the top surface of the adjoining roof covering (see Fig. 8.1b).

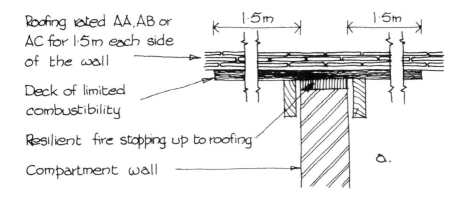

Roofing rated AA, AB or AC for 1.5m each side of the wall

Deck of limited combustibility

Resilient fire stopping up to roofing

Compartment wall

a.

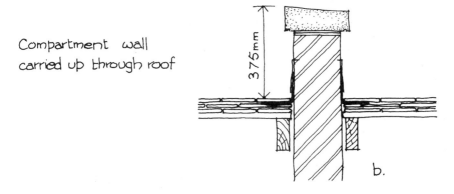

Compartment wall carried up through roof

375mm

b.

Fig. 8.1 Junction of compartment wall and roofing

8.4.2 Openings in fire-resisting walls

Openings in fire-resisting walls that are also compartment walls should be limited to fire-resisting doors, constructed and fitted to give the same standard of resistance as the wall (see Chapter 9), or an opening that allows the passage of a pipe, a ventilation duct or a flue pipe duct. Such openings must be fire stopped or constructed as described in Sections 8.4.3 and 8.4.4.

If the compartment wall separates one or more buildings or the parts of a building in separate tenancies, the only door permitted is one that is needed to provide a means of escape in case of fire.

8.4.3 Fire stopping

Pipes that pass through a compartment wall can be sealed with any proprietary system that has been shown to maintain the fire resistance of the wall; fire stopped, if the diameter is limited as shown below, or sleeved.

With a proprietary sealing system, the diameter of the pipe is not restricted, if it is sleeved the maximum nominal internal diameter permitted is 160 mm and if it is to be fire stopped the maximum internal diameter depends on the material of the pipe as follows:

Any non-combustible material	160 mm
A stack pipe in lead, aluminium alloy, PVC or fibre cement encased in a fire-resisting duct	160 mm
A branch pipe in lead, aluminium alloy, PVC or fibre cement encased in a fire-resisting duct	110 mm
A stack pipe or branch pipe as specified above but not in a duct	40 mm
Any pipe other than those above	40 mm

The fire stopping may be any proprietary material or cement mortar; gypsum-based plaster; cement or gypsum-based vermiculite/perlite mixes; glass fibre, crushed rock, blast furnace slag or ceramic-based products (with or without resin binders), and intumescent mastics.

If a sleeve pipe is to be used it must be of non-combustible materials, such as steel or iron, it must project not less than 1 m on each side of the wall, it must be in contact with the service pipe and the opening through the wall must be as small as possible with fire stopping inserted between the structure and the sleeve pipe.

Where the service pipe is to be encased in a fire-resisting duct, all the internal surfaces, including the wall face(s), are to have a spread of flame classification of Class O. The casing itself should:

- have a fire resistance of not less than 30 minutes, including any access panels
- not be of sheet metal
- not have any access panels opening into a circulation space
- be imperforate except for any openings for a pipe or access panel
- have the openings for pipes as small as possible and with fire stopping run around between the pipe and the structure of the duct

8.4.4 Flues and ducts through fire-resisting and compartment walls

Where a flue or a ventilating duct from an appliance or a chimney containing a number of such flues or ducts passes through, or is built into a fire-resisting wall, each wall of the flue, duct or chimney should have a fire resistance equal to at least half that required for the main wall so that the standard of fire resistance is maintained (see Fig. 8.2).

8.4.5 Cavity barriers

Wherever a cavity is created in a building a path is formed which could lead to the spread of fire throughout the building. As this spread would be concealed

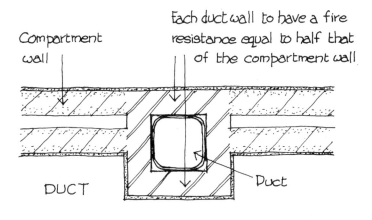

Compartment wall

Each duct wall to have a fire resistance equal to half that of the compartment wall

DUCT

Duct

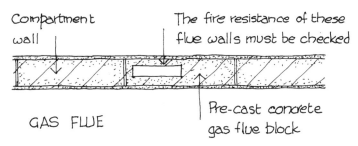

Compartment wall

The fire resistance of these flue walls must be checked

GAS FLUE

Pre-cast concrete gas flue block

Fig. 8.2 Ducts and flues in compartment walls

its effect is far more dangerous than would be the case with a more obvious weakness in the building fabric. There are two principal positions for cavity barriers: in line with the edges of partitions and to subdivide the large concealed spaces that can be created above suspended ceilings.

To prevent fires escaping round the edges of fire-resisting constructions, Approved Document B states that cavity barriers must be installed in line with internal walls in public assembly and residential buildings in the following locations:

- Between a fire-resisting wall in any location and an external cavity wall (see Fig. 8.4c), unless the latter is built as shown in Fig 8.3. If the external walls are constructed as shown in Fig 8.3, to resist the effects of fire, there is no need to provide any barriers in the cavity in line with the edges of compartment walls as these would be continuous with the inner leaf.
- In a protected escape route, above any fire-resisting partitions that are not carried up to the underside of the floor or roof covering over (see Fig. 8.4a), unless the ceiling is fire resisting (see Fig. 8.4d) and it extends throughout the whole building, compartment or separated part.

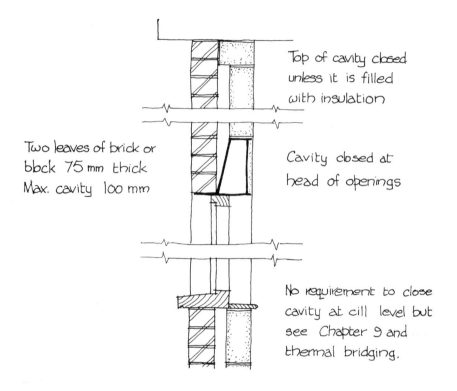

Top of cavity closed unless it is filled with insulation

Two leaves of brick or block 75 mm thick Max. cavity 100 mm

Cavity closed at head of openings

No requirement to close cavity at cill level but see Chapter 9 and thermal bridging.

No combustible material in the cavity except:
A timber lintol, joist end, door or window frame
A pipe, conduit or cable
A dpc, flashing, cavity closer or wall tie
Thermal insulating material

Fig. 8.3 Cavity wall construction not requiring cavity barriers

- Above any bedroom partitions in a residential building that are not carried up to the full height of the storey (or, in the case of the top storey, the underside of the roof covering) unless the ceiling is fire resisting (see Fig 8.4d) and it extends throughout the whole of the building, compartment or separated part.
- In any corridor, above any fire resisting doors that subdivide the corridor to prevent smoke or fire affecting two escape routes simultaneously, unless the doors are part of a fire-resisting partition that subdivides the whole storey.

To prevent the build-up of fire in a large ceiling void or roof cavity its size should be limited by the introduction of cavity barriers spaced at the maximum dimensions given below:

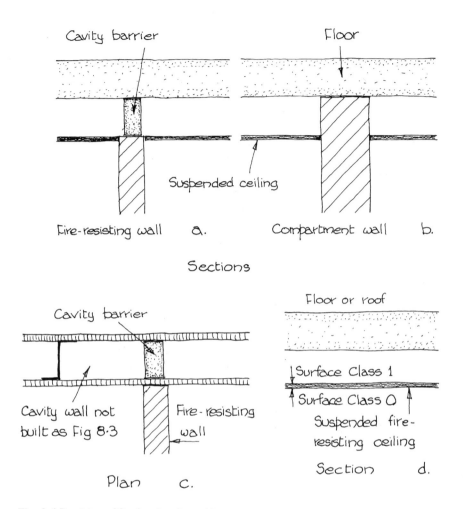

Fig. 8.4 Provision of fire barriers in cavities

Location of cavity	Class of the surface exposed to the cavity	Maximum size in any direction (m)
A roof void to any building	Any class	20
Any other cavity	Class O or Class 1	20
Any other cavity	Any other class	10
Educational building	Class 2 or 3	8
Educational building	Class 4	4
Hospital	Any class	400 m^2 (max. area)

There are a number of situations given in Approved Document B and other legislation where these requirements do not apply:

● Where the cavity extends throughout the building but its extent is no greater than 30 m and the ceiling is fire resisting, as shown in Fig. 8.4d. In addition to the classification of surfaces given in Fig. 8.4, the ceiling should have a 30 minute fire resistance, be imperforate except for an opening as described in Section 8.4.7 below and not be demountable. The requirement that the ceiling be imperforate would apply to any recessed lighting fittings, any terminals to heating or air-conditioning systems run in the ceiling void and any speakers or other equipment fitted into the ceiling. If such installations are required in a fire-resisting ceiling the advice of the manufacturers or the local authority should be sought.

- Where the ceiling void is over a room that is greater than 30 m, but not greater than 40 m in any direction, provided that cavity barriers are installed above the partitions enclosing the room.
- Where the ceiling void is over an undivided area exceeding 40 m in either direction provided that:
  (a) the room and the cavity together are a separate compartment
  (b) an automatic fire detection and alarm system is fitted in the building
  (c) if the cavity is used as a plenum the recommendations about recirculating or distribution systems in BS 5588: Part 9 are followed
  (d) the surface of the ceiling exposed to the cavity is Class O and the supports and fixings in the cavity are non-combustible
  (e) the flame spread rating of any pipe insulation is Class 1
  (f) any electrical wiring is in metal trays or conduit
  (g) any other materials within the cavity are of limited combustibility
- Where the ceiling void is over an operating department in a hospital, cavity barriers are not recommended because of the complexities of the ventilation ductwork and the fact that access to these areas is restricted, they are well supervised when in use and must be enclosed within a fire-resisting construction.

If the internal wall is classed as a compartment wall the use of cavity barriers is not appropriate and the wall must be carried right up through the ceiling void to the compartment floor or roof over (see Fig. 8.4b).

8.4.6 Construction of cavity barriers

A cavity barrier should have a resistance to collapse of 30 minutes and an insulation value of 15 minutes when exposed to fire from either side, unless it is in a stud partition in which case it may be formed of:

- steel at least 0.5 mm thick
- timber at least 38 mm thick
- polythene-sleeved mineral wool compressed into the cavity
- calcium silicate, cement- or gypsum-based boards at least 12.5 mm thick.

Wherever possible a cavity barrier should be tightly fitted to a rigid construction and mechanically fixed in position. If this is not possible, for example fitting to the underside of roof coverings, the junction should be fire stopped. It should also be installed so that its performance is not impaired by movement of the building, the failure or collapse of any fixings, adjoining materials or constructions, or the failure or collapse of any services penetrating the barrier.

8.4.7   Openings in cavity barriers

Any openings in a cavity barrier should be fitted with an automatic fire shutter or a door that should be 30 minute fire resisting and fitted with an automatic self-closing device. It does not need to be able to restrict smoke leakage.

Pipes should be fire stopped as set out in Section 8.4.3 and ducts, unless they are fire resisting, should be fitted with an automatic fire shutter on the line of the cavity barrier.

8.5   Sound insulation

A wall that separates any other building from a dwelling must be capable of keeping the noise of the activities in the property down to a level that will not threaten the health of the occupants of the dwelling and will allow them to sleep, rest and engage in their domestic activities in satisfactory conditions. The sound to be dealt with by the walls is airborne sound. The wording of the guidance given in Approved Document E appears to place the onus for the correct construction of the interposing wall on the designers and builders of the dwelling but, as this may be part of an overall development containing a mix of building uses, the required standards of sound insulation are dealt with in this section. A wall separating a refuse chute from a habitable room or kitchen (or a non-habitable room in an adjoining dwelling) is dealt with in Chapter 14 in Section 14.11.

Approved Document E1 shows four types of wall considered satisfactory, a solid masonry wall, a cavity masonry wall, a masonry wall between isolated panels and a timber-framed wall with absorbency material.

Masonry walls depend on their mass to absorb the sound – being heavy they are not so easily set into vibration – and, as will be seen, the magnitude of their mass determines their sound-insulating property. All masonry walls are composed of a mixture of materials, particularly when the applied finishes are taken into account, and the calculation of their mass must allow for the varying densities of these materials. The method to be used for calculating the mass of a wall is given in Appendix A of Approved Document E and, for normal brickwork coursing at 75 mm, is based on the formula:

$$\text{Mass of } 1\,\text{m}^2 \text{ face area of wall} = \{T(0.79D + 380) + NP\}\,\text{kg/m}^2$$

where: T = thickness of brickwork in metres (unplastered)
D = density of the bricks in kg/m^3 at 3 per cent moisture content
N = number of finished faces (N = 0, 1 or 2)
P = mass of 1 m^2 of wall finish in kg/m^2

The density of the bricks may be taken from a current British Board of Agrément Certificate or a European Technical Approval Certificate related to the material or from the manufacturer's literature, in which case the Building Control Authority may ask for confirmation that the measurement was carried out by an accredited test laboratory. The density of clay bricks and calcium silicate bricks can vary from 1200 to 2400 kg/m^3, common bricks being approximately 1800 kg/m^3.

For the finishes, the mass can be taken from figures given in the Appendix, which assumes a thickness of 13 mm and sets out the following densities:

Cement render	29 kg/m^2
Gypsum plaster	17 kg/m^2
Lightweight plaster	10 kg/m^2
Plasterboard	10 kg/m^2

Other masonry walls, such as concrete block construction, are equally suitable for sound insulation purposes and the Appendix supplies three more formulæ for course heights of 100, 150 and 200 mm, which are:

$$\text{Mass of } 1\,\text{m}^2 \text{ of wall} = \{T(0.86D + 255) + NP\}\,\text{kg/m}^2 \text{ for 100 mm coursing height}$$
$$= \{T(0.92D + 145) + NP\}\,\text{kg/m}^2 \text{ for 150 mm coursing height}$$
$$= \{T(0.93D + 125) + NP\}\,\text{kg/m}^2 \text{ for 200 mm coursing height.}$$

For other coursing heights, the Document states that the formula for the nearest given height can be used.

The constructions given in Fig. 8.5 will achieve the recommended masses that, in the case of the brick wall separating buildings, should be 375 kg/m^2 including the finish, and the concrete blocks and in-situ concrete should be 415 kg/m^2. Care must be taken in the construction of the wall to ensure that all joints are completely filled with mortar and bricks should be laid frog up.

The junction with an external wall shows both a masonry and a timber-framed inner skin with an outer leaf of any construction. Where the inner leaf is of masonry (or the external wall is solid) the walls should be bonded or tied together with ties at not more than 300 mm vertically and the external wall should have a mass of not less than 120 kg/m^2 unless its length is limited by openings which:

- are not less than 1 m high, and
- are on both sides of the separating wall, and
- not more than 700 mm from the face of the separating wall on both sides.

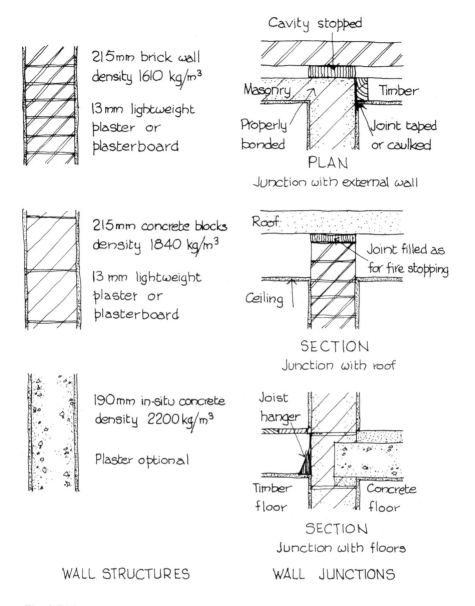

215mm brick wall
density 1610 kg/m³

13mm lightweight
plaster or
plasterboard

215mm concrete blocks
density 1840 kg/m³

13 mm lightweight
plaster or
plasterboard

190mm in-situ concrete
density 2200 kg/m³

Plaster optional

Cavity stopped

Masonry   Timber
Properly
bonded   Joint taped
or caulked

PLAN
Junction with external wall

Roof
Joint filled as
for fire stopping
Ceiling

SECTION
Junction with roof

Joist
hanger

Timber   Concrete
floor   floor

SECTION
Junction with floors

WALL STRUCTURES   WALL JUNCTIONS

Fig. 8.5 Mass masonry sound insulating walls

Where the inner leaf is of timber framing it should be tied to the separating wall with ties at not more than 300 mm apart vertically and the joints should be sealed with tape or caulked.

The cavity wall and the wall with isolated panels depend partly on their mass and partly on the structural isolation between the leaves.

Figure 8.6 shows typical constructions of cavity walls for sound-insulating purposes. The total mass of the brick and the concrete block, including the finish, should be not less than $415\,kg/m^2$ and for the lightweight block wall should be not less than $300\,kg/m^2$. The leaves should be tied together with

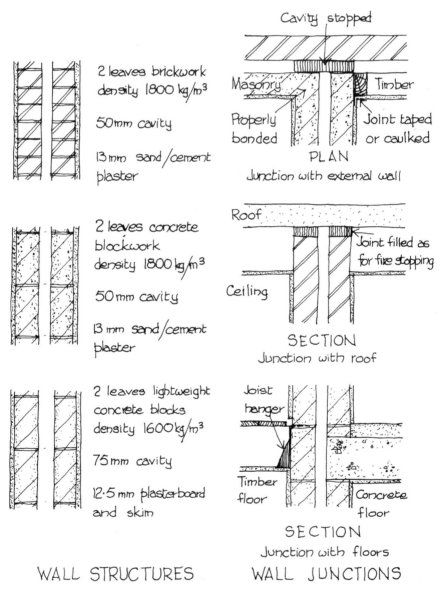

Fig. 8.6 Cavity masonry sound insulating walls

butterfly wire wall ties as required for structural purposes and the cavity kept clear. This means that if the external wall cavity is to be filled with a loose type of thermal insulation material, some means must be used to prevent it from entering the separating wall cavity, such as by the use of cavity batts in the external wall at the junction. The connection with the external wall should be constructed in the same way as a solid sound-insulating wall.

The wall with a masonry core and isolated lining panels, shown in Fig. 8.7, can have a number of combinations of core and panel. Approved Document E defines four cores and two panels.

The four cores are: brick with a mass of $300 \, kg/m^2$, such as $215 \, mm$ of brickwork with a density of $1290 \, kg/m^3$; concrete block, also with a mass of $300 \, kg/m^2$, which can be achieved by a $140 \, mm$ wall of blocks of a density of $2200 \, kg/m^3$; lightweight concrete block with a mass of $160 \, kg/m^2$, which could be a $200 \, mm$ thickness of blocks of a density of $730 \, kg/m^3$; and a cavity core, which can be of any mass but the leaves should be at least $100 \, mm$ thick with a cavity of $50 \, mm$ and tied only with butterfly wire wall ties.

The panels can be either two sheets of plasterboard joined together by a cellular core that should have a mass, including any applied plaster finish, of not less them $18 \, kg/m^2$ and the joints between the panels must be taped, or two sheets of plasterboard with staggered joints. If a supporting framework is used for the double layer of plasterboard each can be $12.5 \, mm$ thick, but if it has no framework the finished thickness must be at least $30 \, mm$.

The junction with an external wall depends on the material of the core. A lightweight block core requires the external wall to be lined with free-standing lining panels similar to the separating wall; where the core is of any of the other three types, the inner skin of the external wall may be plastered or dry lined as shown but must have a total mass of not less than $120 \, kg/m^2$ and be butt jointed to the separating wall with ties at no more than $300 \, mm$ vertically.

Where this type of separating wall supports a suspended timber floor the joists should be carried on joist hangers and solid blocking fixed between them on the line of the isolated panel. Joists running parallel to the wall should be positioned to maintain the air space between the isolated panels and the core. A concrete suspended floor can only be run through the core as shown if its mass is at least $365 \, kg/m^2$ (if it is less, the core wall must run through) and if the core is of cavity form the floor must not bridge the cavity.

Partitions that abut this form of separating wall should not be of masonry construction and should be built as shown with loadbearing partitions secured to the core wall but separated from it by a continuous pad of mineral wool and non-loadbearing partitions fixed to the isolating panels only.

Timber-framed walls rely on isolation and insulation for their sound-insulating properties and a typical section through such a wall is shown in Fig. 8.8. The construction is of two independent stud frames faced with two layers of $15 \, mm$ plasterboard and with a $25 \, mm$ quilt of mineral wool suspended

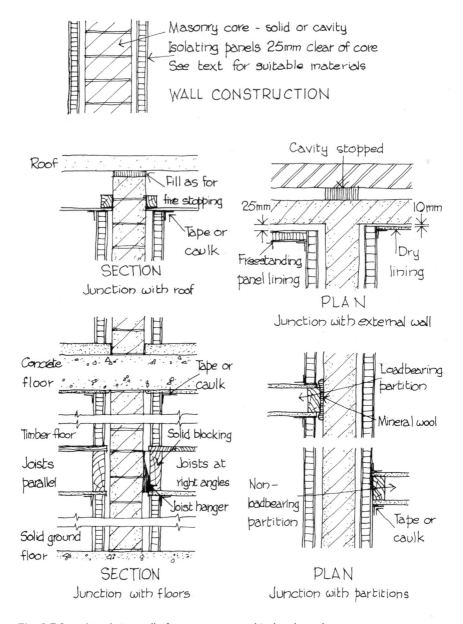

Fig. 8.7 Sound insulating wall of masonry core and isolated panels

between the frames. It is essential with this system that, at no point, is there any firm contact between the two halves of the wall because that would allow a direct transference of sound energy vibrations from one to the other and completely negate the insulating properties.

There are no restrictions on the junction between this type of sound-

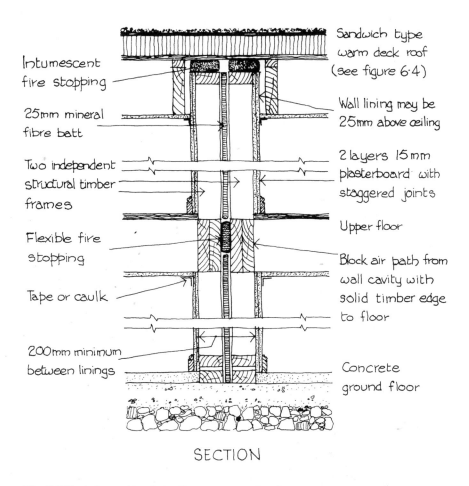

Intumescent fire stopping

25mm mineral fibre batt

Two independent structural timber frames

Flexible fire stopping

Tape or caulk

200mm minimum between linings

Sandwich type warm deck roof (see figure 6·4)

Wall lining may be 25mm above ceiling

2 layers 15mm plasterboard with staggered joints

Upper floor

Block air path from wall cavity with solid timber edge to floor

Concrete ground floor

SECTION

Fig. 8.8 Typical sound insulating fire-resisting timber-framed partition

insulating wall and a timber-framed external wall but if the wall is of cavity masonry construction the cavity must be sealed between the ends of the separating wall and the outer leaf to prevent flanking transmission sound paths being formed.

It should be noted that this construction will also provide a fire resistance of one hour.

8.6 Thermal insulation

The only time that the thermal insulation qualities of an internal wall are of importance is when that wall divides a heated space from an unheated space such as a store or garage. This type of wall is referred to in the Regulations as a semi-exposed wall and Approved Document L states that to comply with the

161

requirements of the Regulations it should have a U value of $0.6\,W/m^2\,K$, if the Elemental Method of demonstration of heat loss is used. Walls with lesser or greater insulation values can be used if either the Calculation Method or the Energy Use Method is employed. A full explanation of these methods is given in Chapter 9.

The Approved Document gives recommendations for wall constructions and the insulation to be inserted to achieve a specific U value. This is summarized in Table 4.10 and illustrated in Fig. 4.10.

8.7 Glazing in internal walls

There are two dangers that should be considered when internal walls are to be glazed: physical injury arising out of the glass being broken, and fire, particularly radiated heat.

With regard to physical injury, any internal walls that contain glazed panels down to floor level are considered to be a hazard unless certain criteria are met. The critical location for such glazing is between finished floor level and 800 mm above the floor. This height is increased to 1500 mm for a distance of 300 mm on each side of a door (and in the door itself, if it is glazed).

The risk can be reduced in these locations if the glazing:

- breaks safely – if it breaks, or
- is robust or in small panes, or
- is permanently protected.

8.7.1 Safe breakage

In practice, safe breakage is concerned with the performance of laminated and toughened glass and is defined in BS 6206: *Specification for impact performance requirements for flat safety glass and safety plastics used in building*: clause 5.3. The requirement of the BS is that safety glass and plastics should either 'not break' or 'break safe'. The latter meaning is that if it does break the impact results in either a small clear opening only with a limit to the size of the detached particles; disintegration into small detached particles or breakage into separate pieces that are neither sharp nor pointed.

It is also a requirement that all panels are marked as safety glass in such a way that it is permanently visible after fixing.

8.7.2 Robustness

Annealed, toughened or laminated safety glass gains its strength through its thickness. The thicknesses shown in Table 8.3 are generally considered to be reasonably safe for ordinary internal glazing except where otherwise recommended.

Table 8.3 Maximum recommended areas of safety glass

Thickness (mm)	Maximum recommended area (m²) of safety glass	
	Annealed glass	Toughened or laminated glass
3	0.2	Not recommended
4	0.3	Not recommended
5	0.45	1
6	0.7	3
10	1.5	6
12	3	7.0

8.7.3 Glazing in small panes

In this context, a 'small pane' may be a single isolated pane of glass or a number of panes contained within glazing bars.

Approved Document N1 recommends that the width of any small pane should be not more than 250 mm and its area not more than 0.5 m². These measurements are to be taken between glazing beads or similar fixings. As can be seen from Table 8.3, the minimum thickness for annealed glass in a small pane of this size is 6 mm except in traditional leaded or copper-light glazing in which 4 mm glass would be acceptable, provided that there is no requirement for fire resistance.

8.7.4 Permanent screen protection

As an alternative to the provision of safety glass or the subdivision of the glazed area into small panes, the glazing can be provided with a suitable form of permanent protection on both sides. The protective screen should be designed and constructed so that:

- it is 800 mm high above the finished floor level
- a sphere of 75 mm diameter cannot come into contact with the glass
- it is sufficiently robust to afford adequate protection
- if the partition containing the glazed area is intended to provide a guard against falling, such as the enclosure at the head of a staircase, the protective screen should not have horizontal bars or any other design feature that would make it climbable (see Fig. 8.9)

8.7.5 Glazing in fire-resisting walls

It is frequently desirable to introduce glazing into the internal partitions in public assembly buildings and it is essential in hospitals and similar buildings

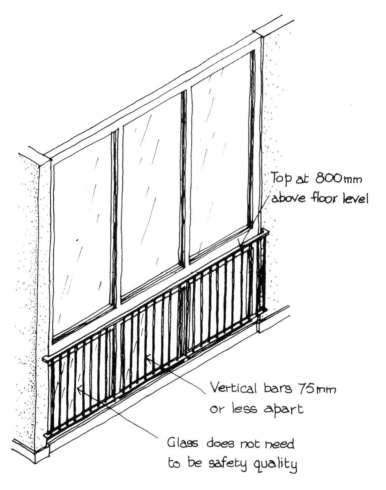

Top at 800mm
above floor level

Vertical bars 75mm
or less apart

Glass does not need
to be safety quality

Fig. 8.9 Permanent protection for wall glazing

where supervision of the occupants is important, but in so doing a fire risk can be created.

Partitions can be glazed so that the glass stays in place in the event of a fire and thus satisfies the integrity criteria given in the Regulations. The hazard is that many of the glasses used permit local heat transmission and radiation through the glass and so are unable to satisfy the insulation criteria. Such heat transmission can constitute a danger to people escaping past the opening and can ignite adjoining combustible material.

Laminated glasses with an intumescent or gel-interlayer can achieve 90 minutes resistance both with respect to integrity and to insulation and can be used without restriction when correctly fitted in a construction designed to resist the effects of fire.

Table 8.4   Limitations on the size of uninsulated glazing in internal partitions

Position of the wall containing the glazing	Permitted glazed area
Between a single protected stairway and a corridor which is not a protected corridor. Between any accommodation and a protected corridor or protected lobby	No glazing permitted
Between more than one protected stairway and a protected corridor. Subdividing a corridor	Unlimited above 0.1 m from the floor
Between a single protected stairway and a protected lobby or protected corridor. Between more than one protected stairway and a corridor that is not a protected corridor	Unlimited above 1.1 m from the floor

(*Source*: Based on Table A4 of Appendix A to Approved Document B (1995 edition))

The traditional annealed wired glass or clear borosilicate glass will achieve the required integrity but its use is restricted because it lacks the required insulation.

In a building where the public can assemble, there is no restriction on the glazing of partition walls from a fire hazard point of view provided that the glazing can achieve a 30 minute rating both for its integrity and for its insulation. If, as is usually the case, this can only be achieved with respect to integrity, the limitations set out in Table 8.4 must apply.

The Firecode for hospitals states that any glazing in a compartment wall must possess the same period of fire resistance in respect of both its integrity and its insulation as that required for the wall itself.

Where the glazing in any building, whether it is used for public assembly, health care, education, or any other type of use, fails to satisfy the insulation criteria, the area that can be glazed is limited to the dimensions shown in Table 8.4.

8.7.6   Manifestation of glazing

It is not always possible to see a large area of internal glazing used as a partition, and a hazard of someone walking into it is thereby created. The risk is at its greatest when the floor is at the same level on both sides of the glazing and anybody might reasonably gain the impression that they can walk directly from one area to the other.

To overcome this danger the regulations require that any uninterrupted areas of glazing are made manifest by the use of broken or solid lines painted on the glass or patterns, company logos or similar designs affixed to it with their centrelines at a height of 1500 mm.

If the glazed area is interrupted with glazing bars, frames or, if it is a door, prominent handles, manifestation is not required. Vertical glazing bars should be spaced at not more than 400 mm. To be of benefit, a horizontal bar should be between 600 and 1500 mm above the floor or, if framing is relied upon to indicate the glazing, it should be substantial.

Chapter 9

WINDOWS, VENTILATION OPENINGS, FIRE DOORS AND ENTRANCES

9.1 The Building Regulations applicable

The Building Regulations applicable to windows, openings for the purpose of ventilation and openings for other purposes in buildings and fire doors where the public are invited to assemble and in buildings with a residential function are:

B4   External fire spread
F1   Means of ventilation
J1   Air supply
L1   Conservation of fuel and power
N1   Glazing – materials and protection

B4 requires that the external walls (including the windows and doors) shall resist the spread of fire over the walls and from one building to another; F1 requires that there shall be adequate means of ventilation provided for people in the building; J1 is concerned that there is an adequate supply of air to heat-producing appliances for combustion and for the efficient working of any chimneys; L1 states that reasonable provision shall be made for the conservation of fuel and power in buildings; and N1 is concerned that glazing with which people are likely to come into contact while in passage about the building, shall either break safely, resist breaking or be protected from impact.

The requirements of the Regulations are not specific in relation to particular building uses and, instead, refer to British Standards and other Governmental publications. In this respect, BS 5588: *Fire precautions in the design, construction and use of buildings*: Part 6: *Code of practice for assembly buildings; Health Technical Memorandum 81* and *Health Technical Memorandum 2025*, both published by NHS Estates; *Activity Data Base*, a computer summary of health care building requirements, produced by NHS Estates; *Fire and the design of educational buildings* produced by the DES and the Home Office *Guide to safety at Sports Grounds* all cover the subject of openings, entrances, exits and ventilation openings and, where relevant, the appropriate recommendations are quoted.

9.2   Thermal insulation

When considering the insulation of a building against heat loss, not only must the walls and roof receive attention, so also must the openings in them and due allowance made for the greater heat loss incurred. The simplest way to deal with this is to set a basic allowance for the area of doors and glazing and in walls and roof and define an average standard U value for the doors, windows and rooflights in those openings. This is the basis for the 'Elemental Method' shown in Section 2 of Approved Document L1. An explanation of 'U value' is given in 4.10.

The area allowances are:

Windows and doors in residential buildings:	30% of the exposed wall area.
Windows and doors in assembly buildings:	40 % of the exposed wall area.
Vehicle access doors in any building:	as required.
Rooflights in any building:	20% of the roof area.

Display windows, shop type of entrance doors and similar glazing may be excluded in calculations for the purposes of assessing heat loss.

The Standard U values are:

Pitched roof up to 70°	$0.25\,\text{W/m}^2\,\text{K}$
Pitched roof over 70°	$0.45\,\text{W/m}^2\,\text{K}$
Pitched roof with no loft space	$0.35\,\text{W/m}^2\,\text{K}$ (Residential buildings)
	$0.45\,\text{W/m}^2\,\text{K}$ (Assembly buildings)
Flat roof	$0.35\,\text{W/m}^2\,\text{K}$ (Residential buildings)
	$0.45\,\text{W/m}^2\,\text{K}$ (Assembly buildings)
Walls exposed to external conditions	$0.45\,\text{W/m}^2\,\text{K}$
Ground floors	
Upper floors exposed to external conditions	$0.45\,\text{W/m}^2\,\text{K}$
Walls and floors exposed to unheated spaces	$0.60\,\text{W/m}^2\,\text{K}$
Windows, general entrance doors and rooflights	$3.30\,\text{W/m}^2\,\text{K}$
Vehicle access and similar large doors	$0.70\,\text{W/m}^2\,\text{K}$

Window and door manufacturers' data are available giving the average U value of their products but, in the absence of such information, the U values shown in Table 9.1, based on Table 7 in Approved Document L1 (1995 edition) can be used.

As can be seen from the table, not all forms of windows, doors and rooflights can achieve the standard U value and some exceed the requirement. This does not mean that they cannot be used but, if they are installed, a corresponding decrease in the permissible area is applied (or increase if the U value is better than the standard). This adjustment is derived from the figures given in Table 9.2, which shows permissible variation in the area of windows, doors and rooflights. In the case of windows and doors, they must be taken

Table 9.1   Indicative U values of doors, windows and rooflights

Element	Glazing	Gap size (mm)	U-values (W/m²K) for a given type of frame			
			Wood	Metal	Thermal break	uPVC
Window	Double glazed	6	3.3	4.2	3.6	3.3
		12	3.0	3.8	3.3	3.0
	Double glazed low-E	6	2.9	3.7	3.1	2.9
		12	2.4	3.2	2.6	2.4
	Double glazed, argon filled	6	3.1	4.0	3.4	3.1
		12	2.9	3.7	3.2	2.9
	Double glazed, low-E, argon filled	6	2.6	3.4	2.8	2.6
		12	2.2	2.9	2.4	2.2
	Triple glazed	6	2.6	3.4	2.9	2.6
		12	2.4	3.2	2.6	2.4
Door	Half double glazed	6	3.1	3.6	3.3	3.1
		12	3.0	3.4	3.2	3.0
	Fully double glazed	6	3.3	4.2	3.6	3.3
		12	3.0	3.8	3.3	3.0
Window/door	Fully single glazed	–	4.7	5.8	5.3	4.7
Door	Solid timber panel	–	3.0	–	–	–
Door	Half single glazed	–	3.7	–	–	–
Rooflight	Double glazed at less than 70° pitch	6	3.6	4.6	4.0	3.6
		12	3.4	4.4	3.8	3.4

(Source: Based on Table 7 of Approved Document L1 (1995 edition))

Table 9.2   Permitted area adjustments related to average *U* value

Average U value (W/m²K)	Permitted area adjustment (% of wall area) for:			Permitted area adjustment for rooflights in all non-residential buildings (% of roof area)
	Windows and doors in places of assembly	Windows and doors in store rooms	Windows and doors in residential buildings	
2.0	74	28	55	37
2.1	69	26	52	35
2.2	65	24	49	33
2.3	62	23	46	31
2.4	58	22	44	29
2.5	56	21	42	28
2.6	53	20	40	27
2.7	51	19	38	25
2.8	49	18	36	24
2.9	47	17	35	23
3.0	45	17	34	22
3.1	43	16	32	22
3.2	41	16	31	21
3.3 (Standard)	**40**	**15**	**30**	**20**
3.4	39	14	29	19
3.5	37	14	28	19
3.6	36	14	27	18
3.7	35	13	26	18
3.8	34	13	26	17
3.9	33	12	25	17
4.0	32	12	24	16
4.1	31	12	23	16
4.2	30	11	23	15
4.3	30	11	22	15
4.4	29	11	22	14
4.5	28	11	21	14
4.6	27	10	21	14
4.7	27	10	20	13
4.8	26	10	20	13
4.9	26	10	19	13
5.0	25	9	19	13

(*Source*: Based on Table 8 of Approved Document L1 (1995 edition))

Notes
1. These adjustments assume that the *U* values of the walls, roof, ground floor, exposed upper floors and semi-exposed upper floors or walls all comply with the standard *U* values.
2. Average *U* value is calculated as shown in section 9.2.

together to find the average U value that can be calculated from the following equation:

$$\text{Average } U \text{value} = \frac{\text{Total rate of heat loss per degree}}{\text{Total area of windows and doors}}$$

Applying this to the hotel building shown in Fig. 9.1, where the area of the windows is $472.8\,\text{m}^2$, or 38.6 per cent of the gross wall area, and that of the entrance doors is $27.2\,\text{m}^2$ or 2.22 per cent, gives:

Element	Area	U value	Rate of heat loss
Windows	$472.8\,\text{m}^2$	$3.0\,\text{W/m}^2\text{K}$	$1418\,\text{W/K}$
Doors	$\underline{27.2\,\text{m}^2}$	$3.8\,\text{W/m}^2\text{K}$	$\underline{103\,\text{W/K}}$
	$500.0\,\text{m}^2$		$1521\,\text{W/K}$

$$\text{Average } U \text{value} = \frac{1521}{500} = 3.042\,\text{W/m}^2\text{K}$$

Referring to Table 9.2, the maximum permissible area for windows and doors with a combined average U value of $3.042\,\text{W/m}^2\text{K}$ is 34 per cent. The actual area is $38.6 + 2.22 = 40.82$ per cent and must, to comply with the requirements of the Elemental Method, be reduced. Compensatory improvements could be made in the insulation standards elsewhere in the building but this cannot be demonstrated using the Elemental Method.

There are two other methods given in Approved Document L1 as alternatives to using the Elemental Method to demonstrate compliance with the requirements of the Building Regulations. These are the Calculation Method and the Energy Use Method. Either of these methods allow the selection of exposed wall, floor or roof systems or windows, doors and rooflights with a U value worse than the standard. However, as a general rule, this should not be poorer than $0.7\,\text{W/m}^2\text{K}$.

The Calculation Method allows greater flexibility in the design and choice of windows, doors and rooflights than the Elemental Method. Its aim is to show that the rate of heat loss from the designed building fabric as a whole will not exceed that of a notional building of the same shape and size designed to comply with the Elemental Method. Clearly, this calculation must take in the heat losses via the walls, ground floor and roof as well as that through the windows, doors and rooflights, all of which can be varied as necessary. Where the U value of the ground floor in the proposed building is better than the standard set, the improved value must also be used for the notional building.

Appendix H to Approved Document L: 1995 Edition gives an example of the application of this method to the four storey office building. The same method can be applied to the hotel building shown in Fig. 9.1, which was examined above under the Elemental Method.

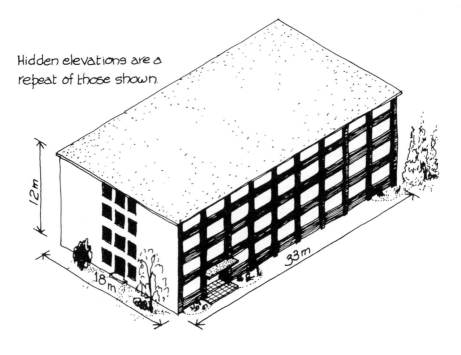

Hidden elevations are a
repeat of those shown.

12 m

18 m

33 m

Notes:

Walls: Brick and block cavity insulated   0.60 W/m²K

Ground floor: Concrete see Table 5.2 (interpolated)   0.315 W/m²K

Roof: Concrete insulated   0.35 W/m²K

Windows: Wood, 6mm double glazing · see Table 9.1   3.30 W/m²K

   Sizes: main 3.0×2.0, end 1.2×2.0

Doors: Metal, double glazed - see Table 9.1   3.8 W/m²K

   Sizes: entrance 3.6×2.83, end 1.2×2.83

Fig. 9.1 54-bed hotel

Dividing the perimeter length of the ground floor by the floor area in this building gives a ratio of 0.17 from which Table 5.1 gives (by linear interpolation) a U value of 0.315 W/m²K, even though it is not insulated (see 5.2.3).

It is the rate of heat loss that must be compared, and that is found by multiplying the area of each element by its respective U value as shown in Fig. 9.1. The rate of heat loss is expressed as the number of watts of heat energy conducted per every degree Kelvin temperature difference between the inside and the outside of the building, and is calculated as follows:

Roof	33×18	
	$= 594.0\,\text{m}^2 \times 0.35\,\text{W/m}^2\,\text{K} =$	208 W/K
Windows	$(3 \times 2 \times 70) + (1.2 \times 2 \times 22)$	
	$= 472.8\,\text{m}^2 \times 3.3\,\text{W/m}^2\,\text{K} =$	1560 W/K
Doors	$(3.6 \times 2.83 \times 2) + (1.2 \times 2.83 \times 2)$	
	$= 27.2\,\text{m}^2 \times 3.8\,\text{W/m}^2\,\text{K} =$	103 W/K
Walls	$[(33 + 18) \times 2 \times 12] - (472.8 + 27.2)$	
	$= 724.0\,\text{m}^2 \times 0.6\,\text{W/m}^2\,\text{K} =$	434 W/K
Floor	33×18	
	$= 594.0\,\text{m}^2 \times 0.315\,\text{W/m}^2\,\text{K} =$	187 W/K

Total rate of heat loss = 2492 W/K

To demonstrate compliance with the Regulation requirements, the proposed building must be compared to a notional building constructed to the standards set out in Approved Document L. This assumes that there are rooflights, with a U value of 3.3 W/m^2 K, to the extent of 20 per cent of the roof area; the windows have a U value of 3.3 W/m^2 K and do not exceed 30 per cent of the wall area; the roof and the walls each possess a U value of 0.45 W/m^2 K and the ground floor will have a U value of either 0.45 W/m^2 K or the same value as the proposed building, if that is better (which is the case in this example). The total heat loss of the equivalent notional building can now be calculated as follows:

Rooflights	$33 \times 18 \times 20\%$	
	$= 118.8\,\text{m}^2 \times 3.3\,\text{W/m}^2\,\text{K} =$	392 W/K
Roof	$(33 \times 18) - 118.8$	
	$= 475.2\,\text{m}^2 \times 0.45\,\text{W/m}^2\,\text{K} =$	214 W/K
Windows and doors	$(33 + 18) \times 2 \times 12 \times 30\%$	
	$= 367.2\,\text{m}^2 \times 3.3\,\text{W/m}^2\,\text{K} =$	1212 W/K
Walls	$[(33 + 18) \times 2 \times 12] - 367.2$	
	$= 856.8\,\text{m}^2 \times 0.45\,\text{W/m}^2\,\text{K} =$	386 W/K
Floor	33×18	
	$= 594.0\,\text{m}^2 \times 0.315\,\text{W/m}^2\,\text{K} =$	187 W/K

Total rate of heat loss = 2391 W/K

As can be seen, the total rate of heat loss of the proposed building is more than that of the notional building (2492 W/K as opposed to 2391 W/K) and therefore the variations from the standard U values are not acceptable. To correct this, either the design must be changed or the specification of the elements must be revised to improve their thermal conductivity. Since windows are the largest contributors to the rate of heat loss, this is the best element to examine first. By referring to Table 9.1, it can be seen that simply by increasing the gap in the double glazing units from 6 mm to 12 mm, the U value goes down by 0.3 to 3.0 W/m^2 K. Applying this better insulation value to the first rate of heat loss calculation reduces the rate of heat loss of the windows to 1418 W/K and the total rate of heat loss of the proposed building to 2350 W/K. Since this is less than the equivalent notional building, the proposals are now acceptable.

The Energy Use Method allows complete freedom of design. As with the Calculation Method, it compares the proposed building to a notional building of the same size and shape but, in this case, the calculation can take into account the employment of any valid energy-saving measures, including useful heat gains from the occupation of the building and the equipment used within it as well as any solar gains. The aim of the Energy Use Method is similar to the Calculation Method but, in this case, it endeavours to show that the total annual energy use of the proposed building would be no more than that of the equivalent notional building.

For buildings that are to be naturally ventilated, an acceptable method of demonstrating compliance with the requirements of the Regulations is given in the CIBSE publication *Building Energy Code* 1981, *Part 2a* (worksheets 1a to 1e).

9.2.1 Thermal bridging around openings

The construction around the perimeter of door and window openings can, in the case of walls or enclosing structures containing cavities that must be closed, lead to an increased conductivity and a consequent thermal bridge. Such thermal bridges significantly affect the energy consumption of the building and create potential local condensation problems.

Figure 9.2 shows the constructional methods recommended in Approved Document L1 that, if followed, will satisfy the requirement. It should be noted that, in all cases, the details involve setting the frame further back in the wall than has been the practice in recent years and makes it necessary to fit a sub-sill.

An alternative to adopting one of these constructions is to show, by calculation, that the edge details will give a satisfactory performance. This can be done either by the procedure given in BRE IP 12/94: *Assessing condensation risks and heat losses at thermal bridges around openings*, or by adopting the procedure given in Appendix D to the Approved Document.

This latter procedure aims to assess the minimum thermal resistance between the inside and outside surfaces at the edges of openings by adding up the thermal resistance of materials along the shortest path. If this resistance is less than $0.20\,\text{m}^2\,\text{K/W}$ (units of thermal resistance are expressed in this form to distinguish them from U value units expressed as $\text{W/m}^2\,\text{K}$) there would be an unacceptable risk of condensation forming at the window reveal and consequential mould growth. This resistance level is also too low to satisfy the thermal insulation requirements unless there is a compensating improvement in the insulation of another element. The amount of compensation necessary would be indicated by taking the increased heat loss around openings into account in any calculations of the building's thermal performance. Appendix D to Approved Document L states that the rate of heat loss in watts

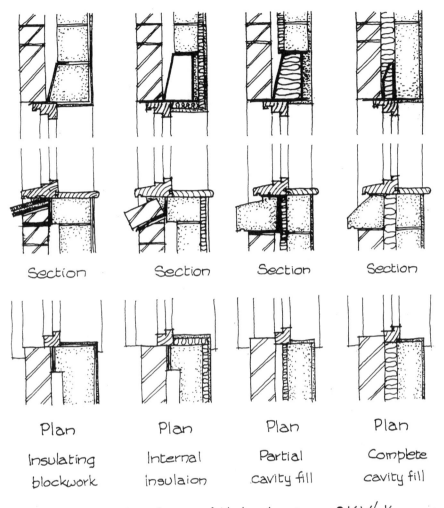

Section   Section   Section   Section

Plan   Plan   Plan   Plan

Insulating   Internal   Partial   Complete
blockwork   insulaion   cavity fill   cavity fill

Notes: Thermal conductivity of blockwork not over 0.16 W/mK
Frame overlap onto blockwork:
     30 mm for dry-lining
     55 mm for lightweight plaster
Steel lintels finished with min 15 mm
lightweight plaster or dry-lined.

Fig. 9.2 Constructions to limit thermal bridging at openings

from this source should be taken as 0.3 times the total length of opening surrounds.

Alternatively, the edge detail could be changed to ensure that the resistance along the shortest path is better than 0.45 m² K/W, in which case the thermal bridging effect can be ignored. Figure 9.3 shows the application of this procedure.

Metal lintels present a particular case in that the thin metal is a good conductor and increases the risk of condensation and mould growth. Addi-

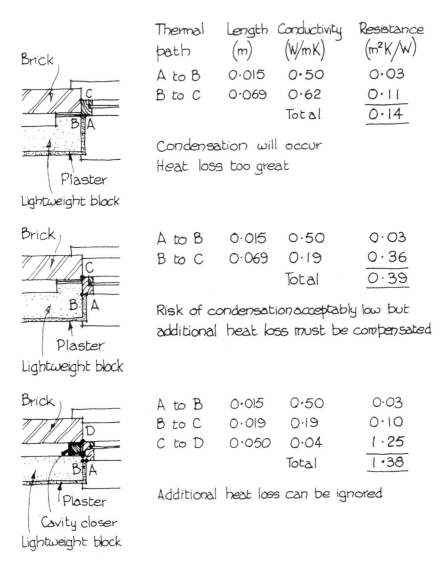

Thermal path	Length (m)	Conductivity (W/mK)	Resistance (m²K/W)
A to B	0·015	0·50	0·03
B to C	0·069	0·62	0·11
		Total	0·14

Condensation will occur
Heat loss too great

A to B	0·015	0·50	0·03
B to C	0·069	0·19	0·36
		Total	0·39

Risk of condensation acceptably low but additional heat loss must be compensated

A to B	0·015	0·50	0·03
B to C	0·019	0·19	0·10
C to D	0·050	0·04	1·25
		Total	1·38

Additional heat loss can be ignored

Fig. 9.3 Thermal bridging around openings

tional, more complex, calculations are needed to check this, as shown in BRE IP 12/94.

9.3 Ventilation

The requirements of Regulation F1 can be met by natural ventilation provided that it achieves the four points set out below:

1. To extract moisture from areas of high humidity.
2. To extract pollutants that are a hazard to health before they become widespread. Such pollutants can arise through processes carried on in the building and in rest rooms where smoking is permitted.
3. To achieve occasional rapid ventilation for the dilution of pollutants and of moisture likely to produce condensation in occupiable rooms and sanitary accommodation.
4. To ensure a minimum supply of fresh air over long periods for the benefit of occupants of the building.

Alternatively, mechanical ventilation or air-conditioning systems are acceptable provided that they are designed, installed and commissioned so that they achieve points 1 to 4 above, operate in a way that is not detrimental to the health of the people in the building and facilities are made for necessary maintenance. In their *Building Bulletin 7*, the Department for Education states that 'where ducting or the ceiling void is used to carry air into more than one space, it is essential that the plant is shut down by the action of smoke detectors as soon as any combustion gas enters the system, or by the operation of the fire alarm'.

Regulation J1 also requires that there is an adequate supply of air to heat-producing appliances for the purposes of combustion and efficiency in the working of any chimney.

The smallest dimension of any ventilation opening should be not less than 8 mm for square or circular holes and 5 mm for slots to minimize resistance to the flow of air. This does not apply to a screen, fascia, baffle, etc.

9.3.1 Definitions

To be clear in this context what is meant by the terms used, the following definitions have been adopted:

Ventilation opening: Any means of ventilation, permanent or closeable, which opens directly to external air. This includes the openable parts of a window, an external door, a louvre, an air brick, or a window trickle ventilator.

Occupiable room: Any room occupied by people. It does not include sanitary accommodation, circulation spaces, plant rooms for services or store rooms.

Sanitary accommodation:	Any room containing a WC or urinal. Sanitary accommodation containing one or more cubicles counts as one space if there is a free circulation of air throughout the space.
Passive stack ventilation:	A ventilating system using ducts running up from the ceilings to terminals on the roof which operate by a combination of the natural stack effect (whereby air moves up the duct due to the difference in temperature between inside and outside air) and the effect of the wind passing over the roof of the building.
Kitchenette:	This term is used in this chapter to describe the small food preparation area found in the staff rooms of many premises where employees can make hot drinks or prepare their own food and where the appliances are of a domestic nature.

9.3.2 Natural ventilation

Satisfactory ventilation of most rooms located on an outside wall, except in very large and complex buildings, can be achieved by ventilation openings and openable windows, assisted, in some cases, by a certain amount of extract ventilation by mechanical means. Table 9.3 sets out the requirements for rapid ventilation openings, background ventilation and extract ventilation as required by Approved Document F1, Table 9.4 gives an extract of the ventilation values recommended by the Chartered Institute of Building Services for a variety of building uses and Table 9.5 sets out the ventilation requirements in health care buildings as shown in Health Technical Memorandum 2025. There are many other spaces and rooms in a hospital requiring very precise ventilation provisions, the detail of which is highly specialized and is dealt with fully in the *Activity Data Base* published by NHS Estates.

The rapid ventilation can be provided by one or more windows or external doors with a total area as indicated in Table 9.3 and with some part at least 1.75 m above floor level. Background ventilation can be by trickle ventilators designed to prevent rain getting in, fitted into or above window or door frames or by air bricks. These should be adjustable and fixed at least 1.75 m above floor level to avoid cold draughts. Extract fans can be operated manually or automatically by sensor or controller.

9.3.3 Mechanical ventilation

In occupiable rooms where a 'no smoking' rule applies, mechanical ventilation must be provided which will achieve a rate of not less than 8 l/s of fresh air per occupant. If light, occasional smoking is permitted, the rate of ventilation must be doubled to 16 l/s of fresh air per occupant. In heavy

Table 9.3   Ventilation of rooms located on outside walls

Room	Rapid ventilation by opening windows or similar	Background ventilation by trickle ventilators or similar	Extract ventilation fan rates
Occupiable room, including a rest room, or similar where smoking is permitted	$\frac{1}{20}$th of the floor area	4000 mm^2 up to 10 m^2 floor area or 400 mm^2/m^2 over 10 m^2 floor area	Smoking rooms only: local extract ventilation to remove tobacco smoke particles
Kitchenette	An opening window (no minimum size)	4000 mm^2	30.0 l/s adjacent to the hob or 60.0 l/s in the room generally
Bath and shower rooms	An opening window (no minimum size)	4000 mm^2 per bath or shower	15.0 litres per bath or shower
Sanitary accommodation, including washing facilities	$\frac{1}{20}$th of the floor area or mechanical ventilation of 6.0 l/s per WC or three air changes per hour*	4000 mm^2 per WC	–
Common spaces where large numbers of people gather, such as a theatre foyer	$\frac{1}{50}$th of the floor area or mechanical ventilation of 1.0 l/s per m^2 floor area	–	–

(*Source*: Based on Table 2 and paragraphs 2.5a, 2.5b and 2.7a of Approved Document F1 (1995 edition))

* Six air changes per hour required in sanitary accommodation in schools.

Note
In areas in schools where noxious fumes may be generated, additional provisions must be made, possibly by the use of a fume cupboard designed in accordance with DFE Design Note 29.

smoking areas, such as rest rooms, the system should be designed to prevent the recirculation of smoke-contaminated air by extracting it to the outside at the rate of 16 l/s per person.

Any kitchenette areas and all sanitary accommodation located away from external walls should be provided with mechanical ventilation as set out in

Table 9.4 Recommended ventilation standards

Building use	Air changes (per hour)	Building use	Air changes (per hour)
Art gallery, museum	1.00	Hotel	
Assembly hall, lecture hall	0.50	Bedrooms	1.00
		Public rooms	1.00
Church or chapel		Corridors and foyers	1.50
up to 7000 m²	0.50	Law court	1.00
over 7000 m²	0.25		
Vestry	1.00	Library	
Dining or banqueting hall	0.50	Reading room under 4 m high	0.75
		Reading room over 4 m high	0.50
Exhibition hall		Stack room	0.50
small (height under 4 m)	0.50	Police station cells	5.00
large (height over 4 m)	0.25		
Gymnasium	0.75	School or college	
		Classroom	2.00
Hospital		Lecture room	1.00
Corridors	1.00	Studio	1.00
Offices	1.00		
Operating theatre suite	0.50	Sports pavilion dressing room	1.00
Wards and patient areas	2.00	Swimming bath: changing rooms	
Waiting rooms	1.00	and bath hall	0.5

(*Source*: Based on Table A4.12 of the Chartered Institute of Building Services Guide A)

Table 9.5 Ventilation rates in health care buildings

Room description	Room pressure with respect to surroundings	Ventilation rates (air changes per hour)		
		Supply	General Extract	Foul extract
WCs	Negative	–	–	10
Bathroom or shower room	Negative	–	–	6
Laboratory	Negative	To suit the room load		–
Treatment room	Balanced	10	10	–
Staff changing room	Positive	3	–	–
Coffee lounge	Negative	–	3	–
Beverage room	Negative	–	5	–
Dirty linen room	Negative	–	–	10
Clean linen room	Positive	6	–	–

(*Source*: Based on Table 2.1 of Health Technical Memorandum 2025, NHS Estates)

Note: Foul extracts must be via a separate system to general extraction.

Table 9.3. A suitable method of switching of the extract fan would be via the light switch or, alternatively, by a sensor that detects a person entering the room. Additionally, an air inlet to the room should be provided, which, at its crudest, could be a 10 mm gap under the door.

Common spaces where large numbers of people congregate, such as a foyer to a public building, should, if not provided with natural ventilation as shown in Table 9.3, be equipped with mechanical ventilation capable of supplying fresh air at the rate of 1.0 l/s for each square metre of floor area. This requirement does not apply to common spaces used solely or principally for circulation purposes.

9.3.4 Ventilation of commercial kitchens and services plant rooms

Approved document F states that the special ventilation requirements of commercial kitchens can be met by compliance with the guidance given in CIBSE Guide B, Tables B2.3 and B2.11. The concern in plant rooms is the provision of emergency ventilation to control the dispersal of gas releases such as a refrigerant leak. Information on this is found in the Health and Safety Executive's Guidance Note EH22: *Ventilation of the workplace*, paragraphs 25 to 27 and in BS 4434: *Specification for safety in the design, construction and installation of refrigeration appliances and systems.*

9.3.5 Ventilation of stage areas

Any stage area must be equipped with high level smoke ventilators equal to 10 per cent of the stage area or mechanical ventilation capable of achieving an equivalent rate of total exhaust airflow. Either form of ventilation must be opened or operated automatically by:

- a fusible device activated at 74 °C, or
- the operation of the sprinkler system, or
- the operation of one of the manual releases that are provided in duplicate, one release on the working side of the stage and one outside the stage area.

Any powered ventilation to the stage area must be automatically cut off when the stage smoke ventilators are opened or operated.

9.3.6 Ventilation of health care buildings

Approved Document F1 states that the ventilation needed for hospitals varies according to the type of accommodation and its function. Detailed requirements are contained in DHSS *Activity Data Base* and in Department of Health Building Notes (HBN) appropriate to each specific department such as HBN 21: *Maternity Departments* or HBN 46: *General medical practice premises.* There is also a comprehensive guidance to the subject in Health Technical Memorandum (HTM) 2025: *Ventilation in health care premises*. The following notes are all taken from the latter publication.

The type and standard of ventilation provision is dependent on:

- the fresh air requirements for habitable rooms
- the extraction of odours, aerosols, gases, vapours, fumes or dust arising from the activity of the particular department
- the need to control or dilute pathogenic material
- the requirements for thermal comfort and control of solar heat gain
- the removal of excessive heat or humidity arising from the equipment used
- the supply of 'make-up air' where local exhaust ventilation is installed
- the combustion requirements of fuel-burning appliances

Both natural and mechanical ventilation have their place, particularly as there is a statutory requirement to provide mechanical ventilation to all enclosed workplaces but, generally, efficient natural ventilation is to be preferred to mechanical extraction wherever possible. In use, it is accepted that natural ventilation is able to give reasonable air distribution for a distance of up to 6 m inwards from the external façade. Beyond this distance a mechanical system would have to be employed. Natural ventilation is created either by wind pressure or the thermo-convective effect of the temperature difference between the inside and the outside of the building. Both these are variable and therefore it is impossible to maintain consistent flow rates. This inconsistency is acceptable in general areas such as staff accommodation, dining rooms and general wards where opening windows should be installed, provided that excessive heat gains or external noise do not preclude their use.

In other areas, a mechanical system would have to be installed, providing extraction only, supply only or a balanced supply and extraction, depending on whether the room pressure should be positive or negative. These pressures and the recommended rates of ventilation are shown in Table 9.5.

The activities within some departments in health care buildings give rise to the need for local exhaust ventilation. This is a statutory requirement under the Control of Substances Hazardous to Health (COSHH) regulations wherever the escape of chemicals, toxic fumes, biological material or quantities of dust would present a hazard to the occupants of a general area.

Mechanical ventilation can vary from an individual extract fan to a full ducted air distribution system. If an individual fan system is used, there should be a 15 to 20 minute run-on time after the fan has been switched off. If a general exhaust system is installed, the replacement air should be filtered and tempered and supplied via a central supply plant adjoining lobbies or corridors to avoid the risk of discomfort that could arise from the ingress of cold air.

Operating rooms require a very specialized cascade system of ventilation to maintain a sterile area around the operating table.

9.3.7   The design of mechanical ventilating or air-conditioning plant

The first requirement is that provision is made to avoid the supply of fresh air being contaminated by pollutants injurious to health, particularly in medical care buildings. Air inlets should not be sited close to a flue, an exhaust ventilation system outlet, an evaporative cooling tower, in an area where vehicles manoeuvre or where polluted air may be drawn into the system.

Care must also be taken to avoid legionella contamination. For guidance on this subject, Approved Document F1 refers to the HSE Guide *The control of legionellosis including legionnaires' disease.* This Guide not only includes design features related to ventilation but also covers a number of other measures.

Satisfactory access for the maintenance of the system is important if it is to be kept in a fully operational condition. A reasonable provision would be to provide access for the purpose of replacing filters and for the cleaning of ductwork.

Adequate space must be provided in the plant room which, if no special provision is necessary (such as the clearance needed to allow tubes to be withdrawn from a boiler) would be satisfied by a 600 mm passageway between items of plant and 1100 mm where routine cleaning is required (see Fig. 9.4). The Approved Document points out that these are minimum requirements and may need to be increased to suit access doors.

Further guidance on more complex situations can be found in the Building Services Research and Information Association Technical Note TN 10/92: *Space allowances for building service distribution systems.*

9.3.8   Commissioning

Approved Document F1 states:

> Where mechanical ventilation and air-conditioning systems are installed to serve floor areas in excess of 200 m² ... the requirement (of Regulation F1) will be satisfied if the building control body is provided with confirmation that the mechanical ventilating systems have been commissioned and tested to demonstrate that they are operating effectively for the purpose of ventilation. A way of demonstrating compliance with the requirements would be to present test reports and commissioning certificates that certify that commissioning and testing have been carried out in accordance with the CIBSE commissioning codes and that the systems perform in accordance with the specification.

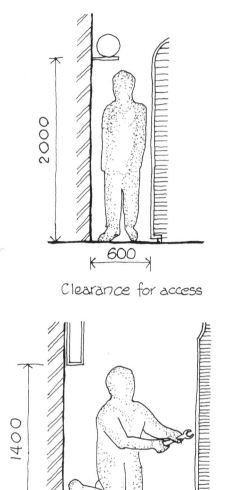

Clearance for access

Clearance for maintenance

Fig. 9.4 Space allowance for plant maintenance

9.3.9 Air supply to heat-producing appliances

As well as ventilation for the comfort of the occupants, ventilation openings are required for the benefit of any solid fuel, gas- or oil-burning appliance unless it is a balanced flue system that allows combustion air in as well as taking the products of combustion out.

Table 9.6 Ventilation openings for the air supply to heat producing appliances (not balanced flue units)

Fuel	Type of appliance	Required total free area of opening
Solid fuel	Open	50% of throat opening
	Any other	550 mm² for each kW of rated output over 5 kW; with a draught stabilizer this must be increased by 300 mm² for each kW of output
Gas	Cooker	Openable window or other means of ventilation of any size; if the room volume is less than 10 m³ a permanent vent of 5000 mm² is needed
	Open flued	450 mm² for each kW of input over 7 kW
Oil	Any	550 mm² for each kW of output over 5 kW

(*Source:* Based on Approved Document J1/2/3)

Different fuels and the type and size of appliance each require a varying amount of air supply and these are summarized in Table 9.6.

9.4 Glazing

Section 8.7 of the previous chapter gave details of the precautions to be taken against glass in vulnerable positions in internal walls causing injury by accidental breakage. The same rules apply to glazing in windows and doors.

The critical location for glazing is within 800 mm of the floor level and would apply to the lower part of a feature window or a glazed door. In this position the glazing, if hit, should either not break or break safely. Alternatively, permanent protection can be provided to prevent the glass being hit.

Prevention of breakage can be achieved by using impact-resistant materials or by restricting the size of the panes to a maximum width of 250 mm and a maximum area of 0.5 m².

Safe breakage is when impact with the glazing results in only a small clear hole and detached particles of limited size, disintegration into small detached particles or shattering into pieces that are neither pointed nor sharp.

The safe size for annealed and laminated safety glazing of various thicknesses is given in Table 8.3 and is reproduced here for convenience as Table 9.7.

Where fire resistance is not a controlling factor, 4 mm glass is acceptable in traditional leaded glazing or copper-lights.

Permanent screen protection is an option more suited to the lower part of windows than to glazed doors but, if provided, removes any need to have safe glazing. The screen should be robust and designed so that a sphere with a diameter of 750 mm cannot pass through at any point to make contact with

Table 9.7   Maximum thickness of safety glass

Thickness (mm)	Maximum recommended area (m²) of safety glass	
	Annealed glass	*Toughened or laminated glass*
3	0.2	Not recommended
4	0.3	Not recommended
5	0.45	I
6	0.7	3
I0	I.5	6
I2	3	7.0

the glass. If the glazing protects against a danger of falling, the screen should be difficult to climb (see Fig. 8.9).

9.4.1   Manifestation of glazing

A further hazard with any large (over 400 mm wide) glazed door panel, or unframed glass door, is the danger that someone may not see it, walk into it and be injured. For this reason, Regulation N requires that such glazing shall incorporate features that make it apparent. With a door, a substantial frame, a horizontal rail between 600 and 1500 mm from the floor or ground level or large door handles or push and pull plates may be considered adequate for this purpose. In the absence of any of these, 'manifestation' must be applied. This can take the form of broken or solid lines, patterns or company logos applied to the glass at a height of 1500 mm above the floor to the centreline of the line, pattern or logo.

9.5   Spread of fire

Chapter 4 deals with external walls and the need for such walls to prevent the spread of fire from one building to another. Windows and doors facing a boundary present a weakness in this fire protection. These, along with any areas of wall that do not meet the fire-resistance standards set out in Table 4.9 are considered to be unprotected areas and must be restricted in their extent, as determined by their distance from the boundary.

While the concern of the Regulations is to prevent the spread of fire from one building to an adjoining property or building, the requirements for the separation of buildings are only applied to the distance from the boundary and not to the distance between buildings. This is to make it possible to calculate the proportion of unprotected area in the wall facing the boundary, regardless of whether, at that moment in time, there is a building on the adjoining site.

By the term 'boundary' the Regulations mean any side of the site that makes an angle with the wall of less than 80°. Usually it also means the actual line of the perimeter of the site; however, it is accepted that the site boundary may adjoin an area where future building works are very unlikely. Spaces such as a road, a railway, a canal or a river are typical of these. In this case it is permitted to extend the boundary to a theoretical line that coincides with the centreline of the feature. Very small openings or small openings that are spaced well apart do not present a hazard and can be ignored. The permitted limits for these small openings are shown in Fig. 9.5. Also disregarded in the assessment of unprotected areas is any part of the external wall of an uncompartmented building that is more than 30 m above mean ground level. No other windows or doors are permitted in a wall that is less than 1.0 m from the boundary.

In walls that are more than 1.0 m from the boundary, the extent of windows, doors and any other unprotected wall areas depends on the purpose of the building, the distance between the boundary and the side of the

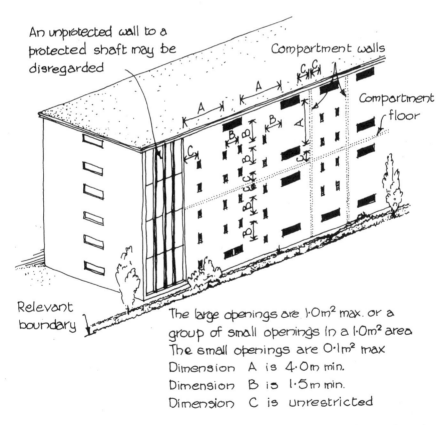

An unprotected wall to a protected shaft may be disregarded

Compartment walls

Compartment floor

Relevant boundary

The large openings are 1·0m² max. or a group of small openings in a 1·0m² area
The small openings are 0·1m² max
Dimension A is 4·0m min.
Dimension B is 1·5m min.
Dimension C is unrestricted

Fig. 9.5 Unprotected areas that may be disregarded when deciding the distance from the boundary

Table 9.8 Permitted unprotected areas in small buildings or compartments

Distance between the side of the building and the boundary (m)	Maximum total percentage of unprotected area in:			
	Residential building and place of assembly		Storage building	
	Fitted with sprinklers	Not fitted with sprinklers	Fitted with sprinklers	Not fitted with sprinklers
1.0	8	8	4	4
2.5	40	20	20	10
5.0	80	40	40	20
7.5	100	60	60	30
10.0	–	80	80	40
12.5	–	100	100	50
15.0	–	–	–	60
17.5	–	–	–	70
20.0	–	–	–	80
22.5	–	–	–	90
25.0	–	–	–	100

(*Source:* Based on Table 16 and paragraph 13.17 of Approved Document B4 (1992 edition))

building and whether a sprinkler system is fitted. The percentage of unprotected area refers to the whole building side, if it is uncompartmented but, if it is divided into fire-resisting compartments, the percentage refers to just the compartment enclosed by that portion of external wall. The maximum permitted unprotected areas are given in Table 9.8 for buildings or compartments up to 10 m high, if this limit is exceeded, the methods set out in the BRE Report: *External fire spread: Building separation and boundary distance* should be used.

The effect of installing a sprinkler system, as can be seen from Table 9.8, is to halve the minimum boundary distance, subject to a minimum distance of 1.0 m being maintained. The sprinklers must meet the relevant recommendations of BS 5306: *Fire extinguishing installations and equipment on premises: Part 2: Specification for sprinkler systems.*

9.6 Fire doors

The important property of a fire-resisting door is its integrity, i.e. its ability to withstand a fire breaking through the door or its frame. The tests for such a fire door assembly are contained in BS 476: Part 22 and are classified by the number of minutes resistance. For instance, a door described as FD30 would prevent flame breaking through for 30 minutes. If the suffix S is added, i.e.

FD30S, the door also restricts the leakage of smoke, which can be as much of a hazard as the fire itself.

Table 9.9 gives the requirements for fire doors in specific locations in terms of their integrity.

Table 9.9   Performance requirements of fire-resisting doors

Function of the wall in which the door is fitted	Minimum fire resistance of the door in terms of integrity
Any compartment wall	As for the wall; if it is on an escape route the door must also restrict smoke leakage
Any compartment wall separating buildings	As for the wall but not less than FD60

Any compartment wall in a hospital which:
  is single storey
  is multi-storeyed up to 12 m high with sprinkler system
  is not described above
	
	FD30S
	FD30S
	
	FD60S

Any sub-compartment wall in a hospital	FD30S

Any wall which forms an enclosure to:
  a protected shaft for a stairway
  a protected lobby or corridor to a stairway
  a lift shaft
  a service shaft
  a boiler room in a school (boiler less than 45 kW)*
  any ancilliary accommodation in an assembly building
  a fire hazard room in a hospital
	
	FD30S
	FD30S
	
	FD30
	FD30 (FD60S in a hospital)
	FD60S (two doors required separated by a ventilated lobby)
	As required for the wall
	
	FD30S

Affords access to an external escape route or stairway	FD30S
Subdivides corridors connecting alternative exits	FD20S
Separates a dead-end corridor from the rest of the corridor	FD20S (FD30S in a school)

(*Source:* Based on Table B1 of Appendix B to Approved Document B (1992 edition), Department of Education's *Building Bulletin 7*, NHS Estates' *Health Technical Memorandum 81* and BS 5588: Part 6)

* If a school boiler room contains a boiler with a capacity in excess of 45 kW output, any doors to the room must be from outside only and not less than 3 m from any other external door.

Note: Unless pressurizing techniques complying with BS 5588: Part 4 are used, smoke-resisting doors (suffix S) should have a leakage rate not exceeding 3 m^3/m/hr (heads and jambs) when tested at 25 Pa under BS 476: Section 31.1.

9.6.1   Self closing fittings

In addition to the door's performance, all fire doors, except cupboard and duct doors that are normally kept shut and locked, should be fitted with an automatic self-closing device.

Where a self-closing device would be considered a hindrance to the occupants of the building in the course of their normal use of the premises, fire doors may be held open by one of the following:

- a fusible link, unless the doorway is a means of escape
- an automatic release mechanism if the door can also be closed manually and it does not lead to either an escape stairway or a fire-fighting stair
- a door closure delay device.

With the exception of lift entrance doors, fire doors should be marked on both sides with an appropriate fire safety sign complying with BS 5499: Part 1 according to whether the door is:

- to be kept locked when not in use
- to be kept closed when not in use
- held open by an automatic release mechanism.

Cupboard and duct doors require fire safety signs on the outside face only.

9.6.2   Ironmongery

The hinges and hardware of a fire door will affect its performance and need to be chosen with care. Guidance is available in a booklet published by the Association of Builders' Hardware Manufacturers in 1983 with the title, *Code of practice for hardware essential to the optimum performance of fire resisting timber doorsets*. Hinges, unless shown to be satisfactory when tested as part of the fire door assembly, should be made from materials with a melting point above 800 °C.

9.6.3   Glazed panels in fire doors

Because they reduce the fire resistance, uninsulated glazed panels in any doors that open into escape routes are limited in size, as shown in Table 9.10. The main hazard to be dealt with in glazed panels is heat transmission. It is possible to glaze a door panel so that it will stay in place in the event of a fire and thus satisfy the integrity requirement, but a high level of radiated heat from the door glazing can constitute a danger to people escaping past the door and can ignite any adjoining combustible material.

Laminated glasses with an intumescent or gel interlayer can achieve a fire resistance of 90 minutes both with respect to integrity and to insulation and can be used without restriction (other than any limitations for reason of

Table 9.10 Maximum sizes of glazed panels in fire-resisting doors

Position of the door	Maximum total glazed area in the parts of the building with access to:	
	A single stairway	More than one stairway
Between the accommodation and:		
a protected stairway	25% of the door area	50% of the door area
a protected lobby	unlimited above 0.1 m from the floor	unlimited above 0.1 m from the floor
a protected corridor forming a dead-end	unlimited above 0.1 m from the floor	unlimited above 0.1 m from the floor
any other corridor	not applicable	unlimited above 0.1 m from the floor
Between a protected stairway and:		
a protected lobby or protected corridor	unlimited above 0.1 m from the floor	unlimited above 0.1 m from the floor
a corridor which is not a protected corridor	25% of the door area	50% of the door area
Subdividing corridors	not applicable	unlimited above 0.1 m from the floor

(*Source*: Based on Table A4 of Appendix A to Approved Document B (1992 edition))

Note: If the protected stairway is also a protected shaft or a fire-fighting stair there may be further restrictions on the use of glazed panels, see Chapter 15.

breakage) when correctly fitted in a door designed to resist the effects of fire.

Any adjoining glazed panels in internal doors are considered as part of the internal wall and are covered in Section 8.7.5.

9.7 Stage safety curtains

Wherever there is a proscenium opening to a stage it is necessary to provide a fire-resisting curtain so that the back stage area can be separated from that occupied by the public in the event of a fire.

BS 5588: Part 6 states that this safety curtain must be lowered once during each performance and it must be entirely non-combustible and able to withstand the effects of a fire for a sufficient period to allow the complete evacuation of the audience. Since the fire-resistance period of the proscenium wall is given as 60 minutes, this would appear to be a reasonable period for the safety curtain.

In addition, the safety curtain must:

- be robust and rigid (which means that it rises straight for its full height)
- be able to withstand damage by falling scenery or properties
- be able to resist the pressure of air likely to be caused by the heat of a fire occurring on the stage
- provide an adequate seal against the penetration of smoke
- close completely within 30 seconds from the time of operation of the curtain release
- have the words 'SAFETY CURTAIN' painted on it conspicuously so as to be clearly visible to the audience
- have the curtain and the guides protected by a hand-operated drencher
- have the hand-release gear provided in duplicate and clearly marked, with one set of gear placed on the working side of the stage and one outside the stage

9.8 Sports Ground entrances and exits

The Home Office *Guide to safety at sports grounds* states that the maximum number of people who should be allowed to enter a sports ground must be based on its holding capacity (the number of people that can be accommodated in the viewing area), the capacity of the entry and exit systems or the capacity of the emergency exit system.

The maximum holding capacity is based on 47 persons per $10\,m^2$, known as the packing density. This density must be reduced where the crush barriers or the conditions of the viewing areas materially deviate from the standards laid down in the Guide (see Chapter 10).

The capacity of the entrance is determined by the number of turnstiles. It is assumed that spectators will pass through a turnstile at the rate of 660 per hour and a simple multiplication of the number of turnstiles will give a maximum capacity for the ground. Conversely, where the number of turnstiles is not fixed, this calculation can be used to determine how many are required for a ground capacity set by other criteria.

The exit capacity is based on a maximum egress time of 8 minutes, this is to be reduced to 2.5 minutes if the exit is also an emergency escape route and there is a risk of rapid fire spread. It can be interpolated between 2.5 and 8 minutes, depending on any active or passive fire safety measures provided. This time is taken as being that in which all the spectators will leave the viewing area, either stands or terraces, and pass into a free-flowing exit system. Once in the exit system spectators should be able to move at the same speed throughout its length. As well as the egress time, two other factors apply: a unit width of 550 mm (the width occupied by a person in the exit route, doorway or gate) and a flow rate of either 40 or 60 persons per minute per unit width. The figure of 40 applies to stairways and exits from stands, 60 applies to terraces and the sports ground areas generally. These values are used in the following formula to find the exit capacity:

$$\text{Exit capacity} = \frac{\text{Width of stairway, passage, door or gate}}{\text{Unit width } (0.55\,\text{m})}$$

$$\times \text{ Flow rate}\,(40 \text{ or } 60) \times \text{Time } (8 \text{ mins})$$

Where practicable, exit gates should be sited near the entrances. Both the exit gates and any doors on the exit route should open outwards in such a way as not to obstruct the exit route and be capable of being secured in an open position before the end of play.

Sliding or roller shutters are not suitable as they do not open when pressure is exerted in the direction of crowd flow and have mechanisms or runways that are vulnerable to jamming.

The Guide also recommends that some of the turnstiles should be reversible or, preferably pass doors should be provided to permit the ejection of anyone whose presence is unwanted or to enable someone to leave the ground at any time. Pass doors should be arranged to operate so that only one person can pass at a time.

Chapter 10

STAIRCASES, RAMPS, TERRACED SEATING AND VIEWING SLOPES

| 10.1   The Building Regulation applicable |

All these aspects are important in a building for the purpose of public assembly, and stairways and ramps are relevant to those used for residential purposes. The standards required for them are laid down in the following Regulations:

B1   Means of escape
K1   Stairs and ramps
K2   Protection from falling
M2   Access and use (by disabled people)
M4   Audience or spectator seating

Regulation B1 requires that the building shall be designed and constructed so that there are means of escape from the building in case of fire, to a place of safety outside the building, capable of being safely and effectively used at all material times; K1 states that stairs, ladders and ramps shall offer safety to users moving between levels of the building; K2 is concerned that stairs, ramps, floors and balconies, and any roof to which people normally have access, shall be guarded with barriers where they are necessary to protect users from the risk of falling; M2 requires that reasonable provision has to be made for disabled people to gain access to and use the building; and M4 stipulates that, if the building contains audience or spectator seating, reasonable provision has to be made to accommodate disabled people.

The means of escape requirements include the provision of staircases that are safe when used by persons who may not be familiar with the building. The whole subject of means of escape is dealt with in Chapter 15. The staircases covered in this chapter are intended solely as a means of access, although, as is mentioned below, in some buildings all stairways should be regarded as a means of escape.

The requirements of Regulations M2 and M4 apply to any of the buildings covered in the description of residential or place of assembly and include, among other provisions, both the means of getting into the building and the design of internal stairs providing access by persons with disabilities and the arrangement of suitable accommodation within a seating area. Where relevant, the standards laid down in Approved Documents M2 and M4 on these subjects are included in this chapter, but the general subject of the measures required to be provided for the benefit of disabled people is covered in Chapter 16.

Approved Document K1 refers to BS 5588: *Fire precautions in the design, construction and use of buildings: Part 6: Code of practice for places of assembly*, and *Guide to Safety at Sports Grounds*, Home Office 1990. Where relevant, the recommendations contained in these documents are included in this chapter.

10.2 The application of Regulations K1 and K2

The requirements of the Regulations apply at any position where there is a difference in floor level in excess of 380 mm or a flight of two or more risers, and cover stairs, ramps and ladders that form part of the building.

The requirements do not apply to means of access outside the building unless the access is part of the building; for example, the requirements do not apply to steps in the length of paths leading to the entrance doors but they do apply to any steps at the entrance doors provided that there are at least two steps or the steps rise more than 380 mm. Nor do they apply to stairs, ramps or ladders that provide access solely for the purposes of maintenance.

10.3 Definitions

To avoid any ambiguity, Approved Document K1 sets out the following meaning to terms used in connection with staircases:

- *Containment*: A barrier that prevents people falling from one floor to the storey below.
- *Flight*: The part of a stair or ramp between landings that has a continuous series of steps or a continuous slope.
- *Going*: The horizontal dimension from front to back of a tread less any overlap with the next tread above.
- *Helical stair*: A stair that describes a helix round a central void.
- *Ladder*: A means of access to another level formed by a series of rungs or narrow treads on which a person normally ascends or descends facing the ladder.
- *Ramp*: A slope of over 1 in 20 designed to take a person from one level to another.
- *Rise*: The height between consecutive treads.
- *Spiral stair*: A stair that describes a helix round a central column.

- *Tapered tread*: A step in which the nosing is not parallel to the nosing of the step or landing above.

Note, particularly, the distinction between a helical stair and a spiral stair.

10.4 Straight flights

It is important that all the steps in a flight of stairs and in each flight where the stair provides access to several floors have exactly the same rise and going and all treads must be level.

In a stair in a building where a substantial number of people will gather, the rise and going should be between the following limits:

- Rise: 135 mm minimum to 180 mm maximum (170 mm maximum if the stair provides access for people with disabilities).
- Going: 280 mm minimum to 340 mm maximum (250 mm minimum if the floor area is less than $100\,m^2$).

In any other building, including sports grounds, the dimension of the rise and going should fall between the following limits:

- Rise: 150 mm minimum to 190 mm maximum (170 mm maximum if the stair is intended for use by disabled people).
- Going: 250 mm minimum to 320 mm maximum (in sports grounds the minimum going is 280 mm and the preferred going is 305 mm).

In either type of building the following rules apply:

- Twice the rise plus the going: between 550 mm and 700 mm.
- Length of flight: if the stair serves an assembly building or a sports ground, the number of risers should be limited to 16 in any one flight. Stairs in other buildings with more than 36 risers in consecutive flights should make at least one turn change in direction of at least 30°.
- Open risers: nosing of the upper tread must overlap the back edge of the lower tread by at least 16 mm and provision must be made so that it is not possible to pass a 100 mm sphere through between the treads if the stair is likely to be used by a child under the age of 5. Open risers should not be used in a stairway in a sports ground.
- Minimum headroom: 2.0 m above the pitch line or landing, maintained, unobstructed, across the full width of the stair.
- Minimum width: no general recommendation is given in Approved Document K. Approved Document M states that it should not be less than 1000 mm, unobstructed, if disabled people are intended to use it and Approved Document B and other documents referred to require that, if the stair is part of an escape route, the minimum width should be not less than the appropriate dimension given in Table 10.1 or not less than the width of any exit leading to the stair.

    If the stair is in an assembly building and is intended to form a means of

Table 10.1   Minimum widths of stairways

Location of stairways	Maximum number of people served	Minimum width (mm)
In an institutional building	150	1000
In an assembly building where the assembly area served is more than 100 m²	220	1100
In a hospital department where the patient beds or trolleys will not be used	200	1100
	Over 200	1100 plus 275 for every additional 50 persons
In an educational building of two storeys	120	1050
In any other building where the stair will be used for escape	over 220	Width calculated as shown in Chapter 15
In any building where the stair is to be used by disabled people	No limit	1000
Any other stair, if required for escape purposes	50	800

(*Source*: Based on Table 6 of Approved Document B1, Approved Document M2, NHS Estates' *Health Technical Memorandum 81* and Department for Education's *Building Bulletin 7*)

Notes
1. The concept of some stairways being for access purposes and some for escape purposes is not acceptable in a hospital so all stairs must provide the full escape capacity.
2. New educational buildings should always have more than one stairway. The width required when two and three stairways are provided are shown in Table 10.2.

escape, the minimum width may be greater than that given in Table 10.1, depending on the number of people served (see Chapter 15).

The Health Technical Memorandum 81 states that all stairways in a hospital should be treated as escape routes and sized accordingly. *Building Bulletin 7* states that, in an educational building, the minimum width of a stair should be 1050 mm. The Home Office *Guide to Safety at Sports Grounds* gives the minimum width of a stairway in a sports ground as 1100 mm.

- Maximum width: an escape stair should not be more than 1400 mm wide if the total rise is not more than 30 m, unless it is at least 1800 mm with an intermediate handrail. In a divided stair, such as this, the resultant widths each side of the handrail must not be less than the minimum widths given above. (This means that in a school, for instance, one cannot have a stair width between 1800 mm and 2100 mm.) The maximum width for a stairway in a sports ground is 1650 mm.
- Handrail: at one side at least, if the stair is less than 1000 mm wide and at both sides if the width is greater than 1000 mm or in a sports ground. In a public building and one used by disabled people, the handrail must extend

down to include the two bottom steps or, in the case of a sports ground, to a distance of 300 mm beyond the top and bottom step. The height of the handrail must be between 900 mm and 1000 mm above the pitch line. The strength of the handrail in a sports ground must be such that it will withstand a horizontal design force of 3.4 kN/m run if it is at right angles to the direction of crowd flow and 2.2 kN/m run if it is parallel to the flow.

● Guarding: where there are two or more steps a guard up to 900 mm above pitch line must be provided which will be able to withstand the horizontal forces shown in Table 10.2. Where the heights match, a handrail can be the top member of the required guarding containment. The containment or guarding to a stair likely to be used by children under five years old must be designed so that a sphere of 100 mm cannot pass through, to prevent a child getting caught, and the guarding must not be easy to climb. Further details on the design and provision of handrails suitable for use by persons with physical difficulties are given in Chapter 16.

All these rules are set out in Fig. 10.1.

If the stair is intended to be used by disabled persons gaining access from the outside into the building, the following variations and additional rules apply:

● Maximum rise: 150 mm.
● Minimum going: 280 mm.
● Maximum total rise: 1200 mm in any one flight.
● Minimum width: 1000 mm.
● Risers to be closed.
● Nosings to be distinguished by a contrasting bright marking.
● Handrail on each side of the flights and landings, whatever the width,

Table 10.2 Minimum widths of stairways in an educational building with two stairways

Maximum number of occupants on all the upper floors of a building where the number of storeys above the final exit door is:									Minimum width of each stairway (mm)
2	3	4	5	6	7	8	9	10	
260	300	340	390	430	470	510	550	600	1050
290	340	390	450	500	550	600	650	710	1200
330	380	440	510	570	630	690	750	820	1350
360	430	500	580	650	720	790	860	940	1500
390	480	560	650	730	810	890	970	1060	1650

(Source: Based on Department for Education's Building Bulletin 7)

Notes
1. The number of occupants in a storey are to be calculated as shown in Chapter 15.
2. Where there are three stairways, the maximum number of occupants can be found by multiplying the figures given in this table by 1.8.

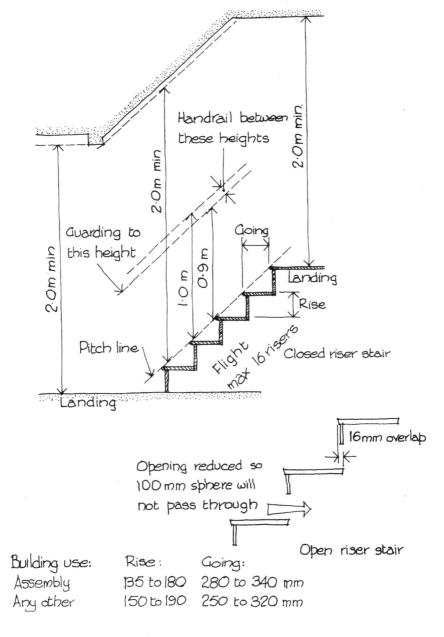

Handrail between these heights

Guarding to this height

Pitch line

Landing

2.0m min

2.0m min

2.0m min

1.0 m

0.9 m

Going

Landing

Rise

Flight max 16 risers

Closed riser stair

16mm overlap

Opening reduced so 100 mm sphere will not pass through

Open riser stair

Building use:	Rise:	Going:
Assembly	135 to 180	280 to 340 mm
Any other	150 to 190	250 to 320 mm

Fig. 10.1 Setting out straight flight stairs

900 mm above the pitch line of the flights and 1000 mm above the edges of the landings.

These rules are set out in Fig. 10.2. It should be noted that the minimum rise

199

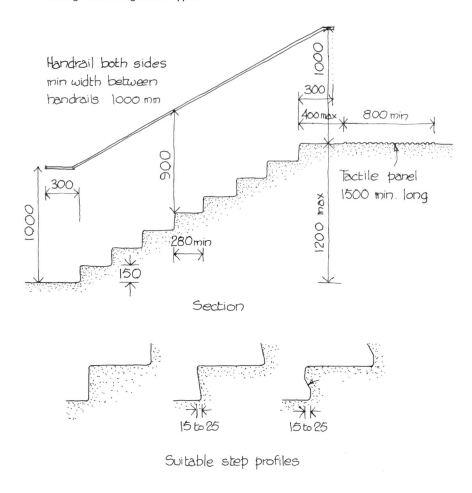

Handrail both sides
min width between
handrails 1000 mm

1000

300

900

280 min

150

1000

300

400 max

800 min

Tactile panel
1500 min. long

1200 max

Section

15 to 25

15 to 25

Suitable step profiles

Fig. 10.2 External access stairs for use by the disabled

for any step is given as 150 mm in Approved Document K1 for any building (except one for public assembly) and the same maximum rise (i.e. 150 mm) is given in Approved Document M2 for a step used by disabled people.

If the stair is intended for use by disabled people within the building the following variations and additional rules apply:

- Maximum rise: 170 mm.
- Minimum going: 250 mm.
- Minimum width: 1000 mm.
- Risers to be closed.
- Nosings to be distinguished by a contrasting bright marking.
- Handrail on each side of the flights and landings, whatever their width, 900 mm above the pitch line of the flights and 1000 mm above the edge of the landings.

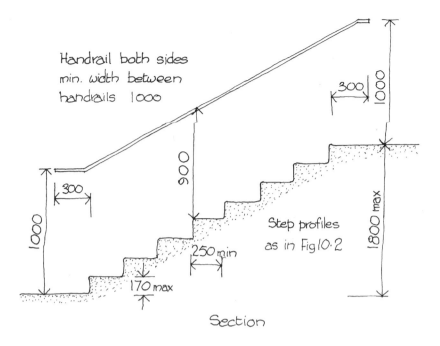

Fig. 10.3 Internal stair for use by the disabled

These rules are set out in Fig. 10.3.

Alternatively, this form of stair can be designed in accordance with the recommendations of BS 5395: *Stairs, ladders and walkways.* Part 1: *Code of practice for the design of straight stairs.*

10.5   Landings

Landings must be provided at the top and bottom of every flight and may include part of the floor of the building. The width and length of each landing must be at least as great as the smallest width of the flight. In addition, if a stair contains more than 36 risers between floors, it must be divided into flights by landings and must change direction by at least 30°.

To afford safe passage, the top landing must be completely clear of any obstructions, the bottom landing and any intermediate landings must be clear of any permanent obstructions but a door may swing across a bottom landing and a cupboard or duct door may swing across an intermediate landing. In both cases, this is only permitted if the door, or doors, leave a clear space of at least 400 mm across the full width of the flight (see Fig. 10.4).

If the stair is intended for use by disabled people, either as an access into the building or as an internal stair, this minimum clear landing length is increased to 1200 mm and, in the case of an external stair, the top landing

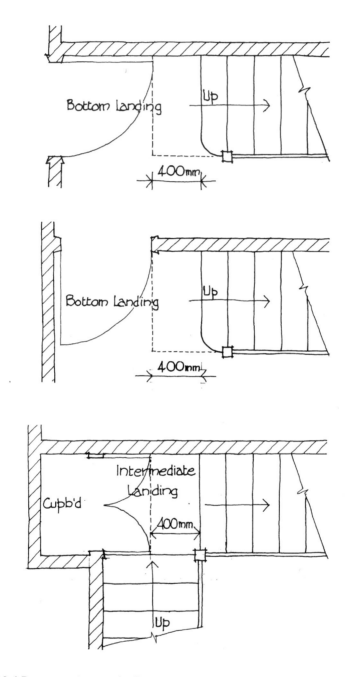

Fig. 10.4 Doors opening over landings

must be marked by a tactile surface to give warning of the change of level (see Chapter 16 for details of tactile paving).

In a hospital, consideration must be given to the requirements for the evacuation of patients, bearing in mind that all stairways in a hospital must be designed for escape purposes. When a stair serves an area where sleeping accommodation is provided, stair widths and landing sizes must be designed to afford enough space to allow patients to be carried out on a mattress. The recommendations of Health Technical Memorandum 81 for landing sizes in relation to stair widths are given in Table 10.3.

The Home Office *Guide to Safety at Sports Grounds* states that the approach to the head of a stairway in a sports ground should be not less than 1100 mm and preferably not more than 3000 mm unless access is totally controlled.

Landings must be level, unless they are formed by the ground at the top or the bottom of the flight, in which case a slope is permitted but it must not exceed 1 in 20.

The minimum headroom height over a landing is 2 m, as shown in Fig. 10.1.

All landings, edges of floors, balconies, galleries, light well, basement area or similar sunken area next to a building, or any area to which people have access must be provided with guarding. Guarding does not need to be provided in loading bays where it would obstruct normal use. The height of the guarding must be 1100 mm except in an assembly building in front of fixed seating where the required height is reduced to 800 mm (but see 10.11 for heights of balconies at the front of tiered seating).

These heights refer to a guard rail with an infill panel below and any wall, parapet, balustrade or similar construction. If the building is likely to be used by children under the age of 5, the design of the guarding must be such that it is impossible to pass a 100 mm sphere through any part of the guarding, nor should it be climbable. For this latter reason, Approved Document K2/3

Table 10.3 Minimum size of landings required to allow mattress evacuation in a hospital

Stair width (mm)	Minimum landing width (mm)	Minimum landing depth (mm)
1300	2800	1850
1400	3000	1750
1500	3200	1550
1600	3400	1450
1700	3600	1400
1800	3800	1350

(*Source:* Based on Table 4 of NHS Estates' *Health Technical Memorandum 81*)

Note: The various widths of stair are determined by how many pedestrians must also use the stair when a mattress evacuation is in progress.

Table 10.4 Horizontal design loads on guarding to stairs, ramps, landings and edges of floors and pavings

Location of the guarding, parapet or balustrade	Uniformly distributed loads		Point load on any part of the infill panel (kN)
	Horizontal linear load (kN/m run)	Distributed load applied to the infill panel (kN/m²)	
In any location in a residential, institutional, educational or public building	0.74	1.00	0.00
Beside any stair, ramp or landing in a theatre, cinema, concert hall, assembly hall, stadium or similar building for public assembly	3.00	1.50	1.50
Along the front edge of any balcony, stand, etc., which has fixed seating within 530 mm* of the guarding	1.00	1.50	1.50
Beside any footway or pavement within the building curtilage, adjacent to an access road, basement or sunken area in a building for public assembly	1.00	1.00	1.00
Beside any pavement or space not less than 3 m wide adjacent to a sunken area such as a light well in a building for public assembly	3.00	1.50	1.50

(*Source*: Based on Diagram 11 of Approved Document K2/3 (1992 edition) and BS 6399: *Loading for buildings*: Part 1: *Code of practice for dead and imposed loads*)

* The dimension of 530 mm is measured from any part of the seating; tip-up seats should be taken as being in their 'up' position.

states that horizontal rails should be avoided. To be effective, the guarding must be capable of resisting the horizontal forces likely to be applied to it. The design loads for this purpose are given in BS 6399 and are shown in Table 10.4.

10. 6   Tapered treads

A tapered tread has a going that varies across its width. Approved Document K1 rules that, in this case, the going shall be measured at the middle if the step is less than 1 m wide or at 270 mm in from each side if it is over 1 m wide. Where the tread or treads are fitted into a rectangular enclosure, the width for this purpose is taken as being from the inner edge of the flight to a curved line connecting the outer edges of steps above and below (see Fig. 10.5).

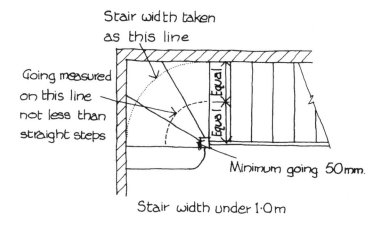

Stair width taken as this line

Going measured on this line not less than straight steps

Equal | Equal

Minimum going 50mm.

Stair width under 1·0m

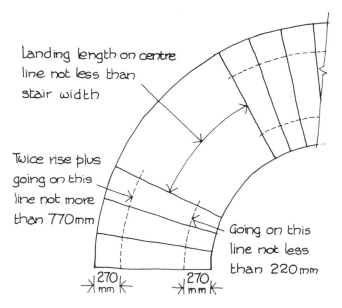

Landing length on centre line not less than stair width

Twice rise plus going on this line not more than 770mm

Going on this line not less than 220mm

270 mm   270 mm

Stair width more than 1·0m

Fig. 10.5 Tapered treads

The maximum and minimum going and rise and the rule that twice the rise plus the going shall be between 550 mm and 700 mm applies to the centreline measurement where the stair is less than 1 m wide and, where it is more, the going at the inner measurement line must not be less than the minimum given and going at the outer line must be not more than the maximum. The going at the inner edge of any tapered tread must not be less than 50 mm.

Tapered treads in a stair to be used as the approach to a building used by

205

disabled people must comply with the minimum going requirement listed above of 280 mm, and this is to be measured along a line 270 mm in from the inner edge of any flight. Approved Document M2, on this subject, does not give any specific figure for the maximum going in a stair to be used by the disabled.

If there are a number of consecutive tapered treads the going must be uniform for each one and if there are straight treads as well as tapered ones, the going of the tapered treads must not be less than that of the straight treads.

Landings may be provided in the length of a series of tapered treads. Their width is measured in the same way as the treads and their length is measured along the centreline no matter what the width (see Fig. 10.5).

Alternatively, any tapered steps designed in accordance with BS 585: *Wood stairs: Part 1: Specification for stairs with closed risers for domestic use, including straight and winder flights and quarter or half landings*, will, as stated in Approved Document K1, offer reasonable safety.

10.7 Spiral and helical stairs

As any spiral or helical stair is a succession of tapered treads, all the rules and recommendations set out above can be used as the basis for a design. Alternatively, the stair can be designed in accordance with BS 5395: *Stairs, ladders and walkways: Part 2: Code of practice for the design of helical and spiral stairs*.

10.8 Ramps for general use

In many respects the standards for ramps for general use, set in Approved Document K1, are the same as for stairs and are summarized below:

- Steepness: the slope should not exceed a rake of 1 in 12 (1 in 10 in sports grounds).
- Width: there is no recommended width except for ramps for use by disabled people (see below).
- Headroom: there should be a clear headroom height of 2 m throughout the length of the ramp and any landings.
- Landings: landings should be provided in the same way as stairs and their length should be at least equal to the width of the ramp.
- Handrails: there is no need to provide a handrail to a ramp that rises less than 600 mm. Ramps under 1000 mm wide should have a handrail on at least one side, ramps over 1000 mm wide should have a handrail both sides. The height of the handrail to the ramp should be between 900 mm and 1000 mm. It should afford a firm grip and, if the heights match, it can form the top of the guarding to the ramp.

- Guarding: all ramps must be guarded by a barrier 900 mm high and any landings must be guarded by a barrier 1100 mm high. The barrier must be designed to withstand the same horizontal forces as a guarding to a landing, as shown in Table 10.4 and, similarly, where the ramp is likely to be used by a child under the age of 5, it should be neither possible to pass a 100 mm sphere through the barrier nor easy to climb.
- Obstructions: ramps must be clear of all obstructions.

10.9   Ramps for disabled people

The design considerations involved in the provision of ramps for disabled people are dealt with in more detail in Chapter 16, however, Approved Document M2 states that the requirements will be satisfied if the ramp complies with the following rules:

- Rake: not more than 1 in 15 if the ramp is not longer than 10 m long or not more than 1 in 12 if the ramp is not more than 5 m long.
- Landings: top and bottom landings 1.2 m long, intermediate landings 1.5 m long, all clear of any door swings.
- Width: surface width 1.2 m, unobstructed width 1.0 m.
- Handrail: both sides of any ramp more than 2.0 m long.
- Kerb: 100 mm high kerb to any open edge of ramp or landing.
- Surface: ramp surface finish to reduce the risk of slipping.
- Steps: where practicable, easy going steps as described above should complement the ramp.

10.10   Fixed access ladders

Approved Document K1 does not give any guidance on the design of fixed ladders. Instead, it states that they should be designed in accordance with BS 5395: *Stairs, ladders and walkways*: Part 3: *Code of practice for the design of industrial stairs, permanent ladders and walkways*, or BS 4211: *Specification for ladders for permanent access to chimneys, other high structures, silos and bins*.

The Approved Document does, however, state that fixed ladders are not suitable as a means of escape for members of the public and should only be used for this purpose where it is impracticable to provide a conventional stair, such as access to a plant room.

BS 4211 deals with the specialized application of ladders. Those fixed permanently to buildings or chimneys, generally externally, to provide a means of access are designated Class A ladders and those fixed to silos and bins, are designated Class B. The Standard gives detailed guidance on suitable materials to be used, the design of the ladder and satisfactory fall arrest systems.

BS 5395: Part 3 gives detailed guidance on the more commonly encountered form of ladder within a building.

The following notes are taken from these Standards.

Sloping ladders are easier and safer than vertical and the Standard defines a ladder as a stair with a pitch greater than 65° and a fixed ladder as a ladder with a pitch in excess of 75° fitted with rungs.

Access to the head of the ladder should be protected by a self-closing gate, no part of the ladder should protrude onto a passageway and the maximum height should not exceed 6 m without a break landing. All landings should be at least 850 mm square and fitted with a handrail and 100 mm high toe plate to all open sides.

The rise, or rung centres, should be between 225 and 255 mm, the width of the ladder should be between 380 and 450 mm inside the strings; these should be robust enough to withstand flexing and be fixed at adequate intervals. There should be a space of at least 230 mm behind each rung to allow foot room and the rungs should be able to withstand a concentrated load of 1.5 kN in any position.

At the top of the ladder the strings should be carried up as handrails, as shown in Fig. 10.6, widening out at platform level to between 600 mm and 700 mm for a height of 1000 mm above the platform. These extended strings should not encroach on the platform or walkway area. Access from the ladder to the platform can be by means of an extension of the platform floor and a flat bar for the top rung (see Fig. 10.6) or additional rungs can be fitted across from the top rung to the platform edge with a maximum space between them of 75 mm.

Tall ladders should be fitted with hoops at 900 mm centres to provide fall arrest protection to anyone using them.

10.11   Terraces, terraced seating and viewing slopes

BS 5588: *Fire precautions in the design, construction and use of buildings*: Part 6: *Code of practice for assembly buildings*, to which Approved Document B refers for a number of standards, also gives guidance on the provision of fixed and temporary seating. In either case the recommendations apply but the latter would not be directly controlled by the Regulations unless it is of the retractable or telescopic type found in a number of multi-purpose halls and sports arenas.

Where the seating is of the retractable or telescopic type, the system should be equipped with locking devices to prevent any movement when the seating is in the extended position. Permanent seating (defined as that where it is improbable that the seats will be removed and the space used for any other purpose) should be securely fixed to the floor.

Approved Document B also refers to the Home Office *Guide to Safety at Sports Grounds* and the recommendations made in the Guide are summarized in this section.

Appoved Document M4 sets out the requirements necessary to meet the

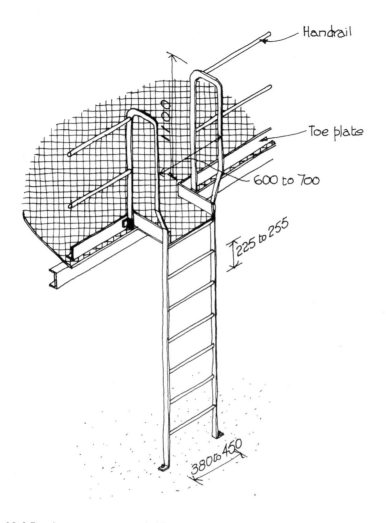

Handrail

Toe plate

600 to 700

225 to 255

380 to 450

Fig. 10.6 Fixed permanent access ladder

Regulation with regard to the provision of accommodation for disabled people within the normal audience area. The standard of the provision is that there should be a number of wheelchair spaces equivalent to 1/100th of the number of fixed seats, subject to a minimum number of six spaces. The disposition and sizes of these spaces are given in Chapter 16.

10.11.1 Seating tiers

In a theatre or similar building the step between one seating tier and the next is usually equal to two steps in the gangway but in all cases, rake of the seating tiers must not exceed 35° and the rise of the steps in the gangways should be between 125 and 190 mm. Each step must be of the same height unless the rake follows a parabolic curve, in which case, the inevitable variation in step must be uniform. The number of gangway steps in any one flight must not exceed 40 if the rake is over 25°. Any runs of more than 40 steps must be broken by cross gangways. Landings must be equal in width to the exit and 1100 mm deep if the exit is from a stepped gangway.

Stepped gangways at the side of theatre seating should be fitted with a handrail at 840 mm above the centre of the steps and not more than 100 mm from the side wall. An open edge to the tiers of seats must be guarded by a balustrade and guard rail at 1100 mm above the centre of the steps. Note that the heights of these handrails and guard rails are measured from the centre of the steps and not the pitch line as in a normal stair.

There should be no change of level between the seatway, i.e. the space between the backs of the seats in one row and the front edge of the seats in the next row (see Fig. 10.7), and the nearest gangway step.

The Standard recommends that the balcony front should be fitted with a coping at least 230 mm wide at a height of 750 mm above the front balcony floor level; if the coping is not 230 mm wide the height of the balcony front should be increased to 790 mm. A guard rail must be fitted, opposite the bottom end of any gangway, equal in width to the gangway and 1100 mm above the front balcony floor (see Fig. 10.7).

In a sports ground, the seating tiers are usually continuous with the gangway steps and should not exceed a rake of 34°. This is the same as the maximum rake of the steps in the gangway as set by the maximum rise combined with the minimum going given in Section 10.11.3. Unlike a theatre, the gangways in a sports ground do not need to contain any intermediate landings but, like a theatre, any side gangways should be provided with a handrail at 840 above the pitch line and projecting not more than 100 mm into the gangway.

10.11.2   Seats

The seats for a closely seated theatre audience, according to BS 5588, must satisfy the criteria for smouldering ignition source 0, flaming ignition source 1 and crib ignition source 5 when tested in accordance with Section 5 of BS 5852.

When fixed in rows, the minimum seatway width is 300 mm, and should increase as the number of seats in the row increases, up to 500 mm when the number of seats exceeds 28 (see Table 10.5). The maximum number of seats in a row with a gangway one side only, is 12 but the maximum number of seats in a row with a gangway both sides is only limited by the travel distance to an exit. Escape routes and travel distances are dealt with in Chapter 15.

The recommendation of the *Guide to Safety at Sports Grounds* is that seats should be securely fixed in position and a space allocated for each person

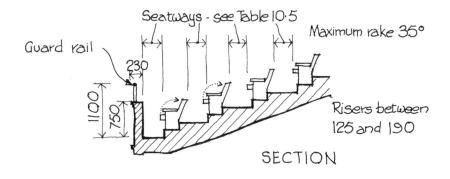

SECTION

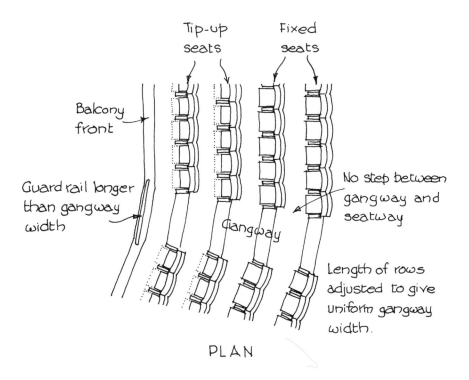

PLAN

Fig. 10.7 Tiered seating

measuring 460 mm wide and 610 mm deep. The width should be increased to 500 mm if the seats are fitted with arms and, for comfort, the dimensions should be further increased to 550 and 760 mm respectively. The seatway width should be at least 305 mm and is included in the depths of spectator space given above. There should not be more than 14 seats to a gangway, so

211

Table 10.5 Maximum number of fixed seats in a row in an auditorium

Seatway width (mm)	Maximum number of seats in a row with:	
	Gangway one end	Gangway both ends
300 to 324	7	14
325 to 349	8	16
350 to 374	9	18
375 to 399	10	20
400 to 424	11	22
425 to 449	12	24
450 to 474	12	26
475 to 499	12	28
500 or more	12	Limited by travel distance

(*Source*: Based on BS 5588: *Fire precautions in the design, construction and use of buildings*: Part 6: *Code of practice for assembly buildings*)

the maximum number in a row with a gangway one end only is 14 and with a gangway both ends is 28.

10.11.3 Gangways

The minimum gangway width, if there are less than 50 people in a theatre audience, is 900 mm; if there are more than 50, the minimum width is 1100 mm; in either case measured clear except for a handrail projecting no more than 100 mm from a wall. In all cases in a sports ground, the minimum width for a gangway is 1100 mm.

The length of rows of different numbers of seats must be adjusted so that their ends are aligned to preserve a uniform gangway width, unless escape is in one direction only, in which case the width of the gangway should increase towards the exit.

The radial gangways should intersect with the transverse gangways as a 'T' junction and these should be offset to ensure a smooth flow towards the exit in a theatre and to discourage spectators in a sports ground from using them for viewing purposes (see Fig. 10.8).

Gangways in sports grounds should be painted with a non-slip paint in a conspicuous colour.

10.11.4 Terraces and viewing slopes

The Home Office *Guide to Safety at Sports Grounds* states that terraces are recognized as presenting a special safety problem due not only to the incidental dangers of spectators standing for a long time on steep, terraced

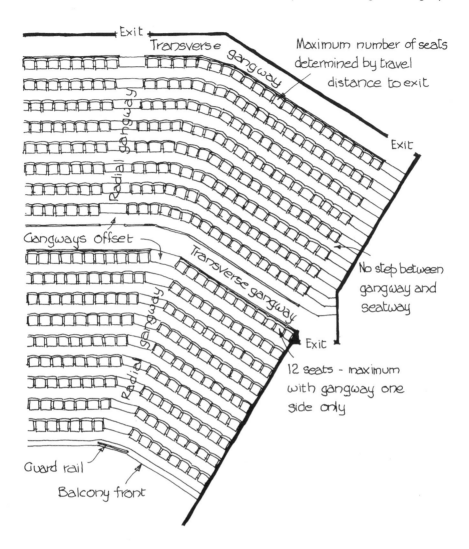

Fig. 10.8 Auditorium layout

slopes but also because of excitement generated and pressures and surges caused by people straining to see the event. An unobstructed view from all parts of the terrace is important to minimize this danger.

The steps of a terrace should have a rise between 75 and 180 mm and a going between 280 and 380 mm, preferably between 355 and 380 mm. If the rise has to exceed 180 mm a crush barrier must be provided at the top of the rise.

The rake of the terrace should not exceed 1 in 2 and excessive variations in the gradient should be avoided.

The surface of the terrace steps should be of a non-slip material and drained to prevent deterioration in wet weather.

All spectators should be within 12 m of a gangway or exit, measured along a line of unobstructed travel from their viewing position. The gangway width should be not less than 1100 mm and linked to the exit system.

Viewing slopes should also be covered with a non-slip surface and their preferred maximum gradient should be 1 in 6 or 9.5°. Viewing slopes with any steeper gradients should be provided with continuous crush barriers from one radial gangway to the next.

The minimum recommended distance between the front of the terrace or viewing slope and the touch line or playing area is 3 m where the terrace gradient is 30° or greater. This should be increased as the gradient reduces so that spectators will be able to obtain a clear view.

10.11.5 Crush barriers

Crush barriers should be provided at the foot of gangways or stairways and, as mentioned above, continuously across viewing slopes between one gangway and the next, or wherever the height of a terrace step exceeds 180 mm. The maximum spacings for the barriers depend on the gradient of the slope and are set out in Table 10.6. The strength requirements are greater for the wider spacing as shown and values can be interpolated between the two limits. The design loads shown in Table 10.6 are static loads applied horizontally to the

Table 10.6 Spacing of sports ground crush barriers

Terrace gradient (degrees)	Maximum horizontal distance (m) between crush barriers designed to withstand horizontal design force (kN/m run)* of:	
	5.0	3.4
0.5	5	3.3
10	4.3	2.9
15	3.8	2.6
20	3.4	2.3
25	3.1	2.1
Gradients in excess of 25 degrees are potentially hazardous and should be avoided		
30	2.9	1.9
Gradients should not exceed 30 degrees		

(*Source*: Based on Table 1 and Table 2 of Home Office *Guide to Safety at Sports Grounds*)

* The design force is to be taken as a static force applied horizontally to the top rail of the crush barrier at right angles to its longitudinal axis.

top member of the barrier at right angles to the longitudinal axis. Barrier foundations should be capable of resisting the overturning moments and the sliding forces induced by the design loads, with a factor of safety of 2.

The height of the barriers should be between 1020 and 1120 mm, with a preferred height of 1100 mm. There should be no sharp projections or edges in the barrier and research has shown that a flat top rail of 100 mm width, fixed vertically, is better than the 50 mm tube used at many sports grounds.

Handrails not forming crush barriers and fixed at right angles to the direction of flow, should be capable of resisting a design load of 3.4 kN/m width applied to the top rail, and handrails parallel to the flow should be able to resist a design load of 2.2 kN/m width.

Chapter 11

HEARTHS FOR SOLID FUEL APPLIANCES, FLUES AND CHIMNEYS

| 11.1   The Building Regulations applicable |

Although this type of installation is never found in some of the building types covered in this book, the range is so large that there could well be some where the provision of a solid fuel appliance is desired. In that case, the Building Regulations covering such an installation are:

B3   Internal fire spread (structure)
J1   Air supply
J2   Discharge of products of combustion
J3   Protection of building

B3 refers, very briefly, to flues in connection with compartment walls and requires that the construction shall inhibit the spread of fire; J1 is concerned that there is an adequate supply of air to an appliance for combustion and for the efficient working of any flue pipe or chimney; J2 states that heat-producing appliances shall have adequate provision for the discharge of the products of combustion to the outside air; and J3 requires that heat-producing appliances and flue pipes shall be so installed, and fireplaces and chimneys so constructed, that the risk of the building catching fire in consequence of their use is reduced to a reasonable level.

The following recommendations are based on Approved Document J1/2/3 and apply to solid fuel appliances only. Gas-fired appliances with a rated input up to 60 kW and oil-fired appliances with a rated output up to 45 kW are dealt with in Chapter 12. The larger installations found in most buildings used for assembly and many residential buildings are the subject of specialist design and construction and are, therefore, not dealt with in this book.

| 11.2   Hearths for free-standing solid fuel appliances |

Any free-standing solid fuel appliance must be placed on a constructional hearth that is of solid non-combustible material at least 125 mm thick (see Fig.

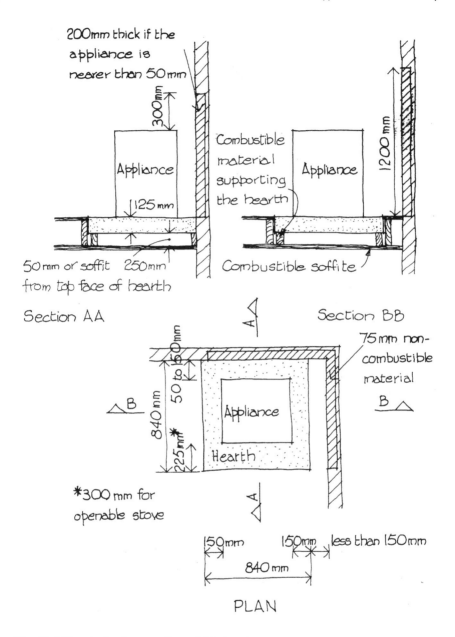

Fig. 11.1 Hearths for free-standing solid fuel appliances

11.1). This thickness can include the thickness of any non-combustible floor under the hearth. The minimum size of the hearth is 840 mm square. This minimum size may need to be increased if the plan size of the boiler is greater than 465 mm deep and 540 mm wide because, in plan, the hearth should project 150 mm each side and behind the fitting and 225 mm in front if it is a closed stove, or 300 mm in front if it is a type that can be used with the front open.

If the hearth is larger than the minimum that this would give, combustible material may be placed on it but no nearer than 150 mm each side and behind the appliance and 225 or 300 mm in front.

Any wall behind or beside the appliance and less than 150 mm from the edge of the hearth, should be of non-combustible material for a height of at least 1.2 m and for a thickness of at least 75 mm. If the wall is between 150 and 50 mm from the appliance, the non-combustible material must extend at least 300 mm above the fitting and, if it is under 50 mm from the appliance, its thickness must be increased to 200 mm.

11.3 Flues and flue pipes

Flue pipes for solid fuel appliances may be made in any of the materials listed below but should only be used to connect the appliance to a chimney. In addition they should not pass through a roof space.

- Cast iron as described in BS 41: *Specification for cast iron spigot and socket flue or smoke pipes and fittings.*
- Mild steel with a wall thickness of at least 3 mm.
- Stainless steel with a wall thickness of at least 1 mm and as described in BS 149: *Steel plate, sheet and strip.* Part 2: *Specification for stainless and heat resisting steel plate, sheet and strip* for Grade 316 S11, 316 S13, 316 S16, 316 S31, 316 S33, or the equivalent Euronorm 88-71 designation.
- Vitreous enamelled steel complying with BS 6999: *Specification for vitreous enamelled low carbon steel flue pipes, other components and accessories for solid fuel burning appliances with a maximum rated output of 45 kW.*

All flue pipes with spigot and socket joints should be fitted with the socket facing upwards.

11.3.1 Sizes of flue pipes

A flue pipe should not be smaller than the outlet of the appliance to which it is connected.

11.3.2 Direction of flue pipes

Flues should be run vertically wherever possible. Horizontal runs should be avoided except in the cases of a back outlet appliance, when the length of flue between the appliance and a chimney should not exceed 150 mm.

If a bend is required in a flue, the angle it makes with the vertical should not exceed 30°.

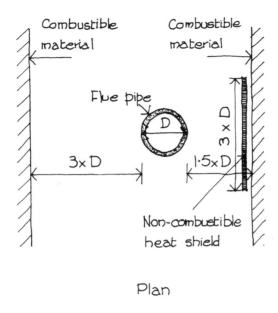

Plan

Fig. 11.2 Separation of solid fuel flue pipe and combustible material

11.3.3 Shielding of flue pipes

A flue pipe from a solid fuel appliance must be separated from anything that might burn by at least 200 mm of non-combustible material or an air space of not less than three times the diameter of the pipe. If a non-combustible shield is provided, equal in width to three times the diameter of the pipe and spaced away from the combustible material by at least 12.5 mm, the air space required can be reduced to one and a half times the diameter of the pipe (see Fig. 11.2).

If the wall or floor is of the compartment type as described in Chapters 6 and 8, the flue pipe must be cased with non-combustible material with at least half the fire resistance specified for the wall or floor.

11.3.4 Outlets from flues

The outlet from a flue contained within a chimney serving a solid fuel appliance must be carefully positioned in relation to the roof line, as shown in Figs 11.3 and 11.4.

As can be seen from Fig. 11.3, the Approved Document does not give a direct height of chimney above a pitched roof, instead it gives a zone outside which the flue outlet must be placed. The effect of this is that the actual minimum height of chimney required is not fixed, except in relation to the ridge, and is determined by the pitch of the roof. The graph in Fig. 11.3 is an interpretation of the Approved Document rule and from it can be read both

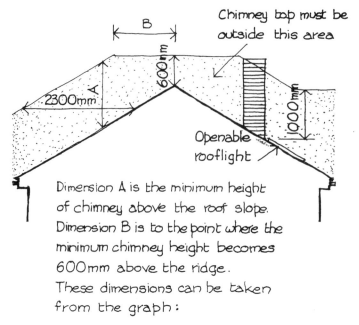

Dimension A is the minimum height
of chimney above the roof slope.
Dimension B is to the point where the
minimum chimney height becomes
600mm above the ridge.
These dimensions can be taken
from the graph:

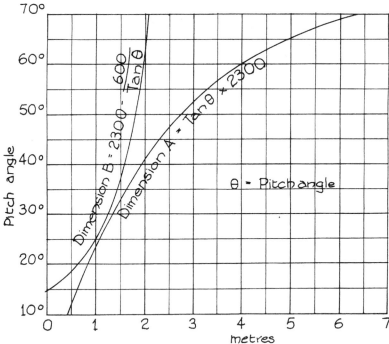

Fig. 11.3 Height of chimney above a pitched roof

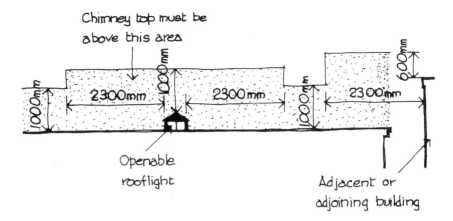

Fig. 11.4 Height of chimney above a flat roof

the direct height and the distance from the ridge to the point where the required height becomes 600 mm above the ridge.

For the frequently used pitch of 35° the minimum height of the flue outlet above the highest point of intersection of the chimney with the roof is 1615 mm, but if it is within 1443 mm of the ridge the top should be 600 mm above the ridge.

The outlet from a flue attached to a pressure jet oil-burning appliance can be terminated anywhere above the roof.

11.4 Chimneys

Chimneys can be built to serve solid fuel, gas-fired or oil-fired appliances and, to a very large extent their construction is the same no matter what the fuel.

11.4.1 Brick and block chimneys

Brick chimneys serving any type of appliance should be lined. The liners should be of one of the types listed below, fitted with the socket or rebate facing upwards to prevent any condensate from running out and jointed with fire-proof mortar. Any space between the liners should be filled with weak mortar or insulating concrete.

Types of liner listed in Approved Document J1/2/3 are:

- Clay flue liners with rebated or socket joints as described in BS 1181: *Specification for clay flue linings and flue terminals.*
- Imperforate clay pipes with socketed joints as described in BS 65: *Specification for vitrified clay pipes, fittings and joints.*
- High alumina cement and kiln-burnt or pumice aggregate pipes with rebated or socketed joints or steel collars around the joints.

A blockwork chimney serving a solid fuel appliance should either be lined as for a brick chimney, or built from refractory materials or a combination of high alumina cement and kiln-burnt or pumice aggregate. Blockwork chimneys for gas- or oil-fired appliances with flue gases cooler than 260 °C can be built with purpose-made flue blocks conforming either to BS 1289: *Flue blocks and masonry terminals for gas appliances*: Part 1: *Specification for precast concrete flue blocks and terminals* or Part 2: *Specification for clay flue blocks and terminals*.

The size of the flue inside a chimney should never be less than the size of the outlet of the appliance connected to it, or the size recommended by the manufacturer of the appliance, or:

- 125 mm for a closed appliance up to 20 kW rated output not burning bituminous coal
- 150 mm for a closed appliance up to 20 kW rated output burning bituminous coal
- 150 mm for a closed appliance from 20 to 30 kW rated output
- 175 mm for a closed appliance from 30 to 45 kW rated output

The walls of a chimney, excluding the liners, should be not less than 100 mm thick unless the wall is between the flue and either another compartment of the same building or another building, in which case it must be 200 mm thick.

Combustible material should be separated from the chimney by either 200 mm from the .flue or 40 mm from the outer face of the chimney or fireplace recess, unless it is a floorboard, skirting, or architrave. Metal fixings in contact with combustible materials should be at least 50 mm from the flue.

11.4.2   Factory-made chimneys

Factory-made insulated chimneys should be constructed and tested to meet the relevant recommendations given in BS 4543: *Factory-made insulated chimneys*: Part 1: *Methods of test for factory-made insulated chimneys*, Part 2: *Specification for chimneys for solid fuel appliances*, or Part 3: *Specification for chimneys with stainless steel flue linings for use with oil fired appliances*, as appropriate. They should be installed in accordance with the manufacturer's instructions.

A factory-made insulated chimney or an insulated metal chimney intended to serve a gas- or oil-fired appliance, should not pass through any part of the building unless it is cased in non-combustible material giving at least half the fire resistance of the compartment wall or floor. Nor should it pass through a cupboard, storage space or roof unless it is surrounded by a non-combustible guard fixed at a distance from the outer surface of the chimney shown to be safe by test in accordance with BS 4543: Part 1. This same safe distance applies to the spacing that must be allowed between the outer wall of a factory-made insulated chimney and any combustible material.

11.4.3 Debris-collecting space

Any chimney to a solid fuel appliance that is not directly over the fitting or, one serving a gas- or oil-fired appliance that is not lined or not constructed of purpose-made flue blocks must be provided with a debris-collecting space. This space must be at the bottom of the chimney, have a volume of at least $0.012\,\text{m}^3$, a depth of at least 250 mm below the point of connection of the appliance and be accessible for the purpose of clearing any debris collected. Access requiring the removal of the fitting is permissible.

11.5 Air supply

Unless the appliance is room sealed, the room or space in which it is sited should have a ventilation opening of the size required for the type and capacity of the appliance. If this opening is to an adjoining room or space then that too must have a ventilation opening of the same size that vents directly to the open air.

Ventilation openings should not be in fire-resisting walls.

The size required for either open or closed appliances is $550\,\text{mm}^2/\text{kW}$ of rated output above 5 kW. If a draught stabilizer is fitted, this must be increased $300\,\text{mm}^2$ for each kilowatt of output.

Because of the ventilation requirements to be observed in a kitchenette (see Chapter 9, Section 9.3), no internal kitchenette (which has to rely entirely on mechanical ventilation) should contain an open-flued appliance as there is a possibility of flue gas spillage.

Natural ventilation only should be provided to any kitchenette containing a solid fuel appliance.

Chapter 12

GAS- AND OIL-FIRED EQUIPMENT, HEATING, HOT WATER AND LIGHTING

12.1 The Building Regulations applicable

The Building Regulations that apply to any of these types of installation in a building to be used for the purpose of assembly or residence are:

G3 Hot water storage
J1 Air supply
J2 Discharge of products of combustion
J3 Protection of buildings
L1 Conservation of fuel and power

G3 requires that unvented hot water systems are safe; J1 sets standards for an adequate supply of air for combustion purposes; J2 states that heat-producing appliances shall have adequate provision for the discharge of the products of combustion to the outside air; J3 requires that heat-producing appliances and flue pipes shall be so installed as to reduce to a reasonable level the risk of the building catching fire in consequence of their use; L1 contains a section setting out acceptable standards of insulation of hot water storage vessels, pipes and ducts and it also requires that in non-domestic buildings where more than $100\,\text{m}^2$ of floor area is artificially lit, the system shall be designed, constructed and controlled so that no more fuel than is reasonable is consumed.

The following recommendations are based on Approved Documents G3, J1/2/3 and L1 and refer to gas- and oil-fired appliances. Solid fuel burning equipment is dealt with in Chapter 11.

12.2 Hearths

Gas- and oil-fired appliances are all contained within themselves and the design of many of them is such that there is no hazard arising from the floor below the appliance becoming overheated. Approved Document J1/2/3 sets out recommendations for hearths suitable to receive these boilers and similar appliances. It has been assumed that the only type of appliance to be encountered in a building of the type covered in this book is a boiler or water heater. If, for a particular reason, say in an elderly person's accommodation, a fuel effect gas appliance is required, the installation should follow either the manufacturer's instructions or the recommendations of BS 5871: *Specification for installation of gas fires, convector heaters fire/back boilers and decorative effect gas appliances.*

12.2.1 Hearths for gas-fired appliances

A hearth should always be provided under a gas-fired appliance unless every part of the flame or incandescent material is at least 225 mm above the floor or the appliance complies with the recommendations of BS 5258: *Safety of domestic gas appliances* or BS 5386: *Specification for gas burning appliances for installation without a hearth.*

In the case of a fitting that includes a back boiler and requires a hearth, its dimensions should be 150 mm larger than the appliance at the sides and back, 225 mm larger in front and either 125 mm thick and solid or 25 mm thick on non-combustible supports 25 mm high (see Fig. 12.1).

Any other appliance, should be installed on a non-combustible hearth of the same dimensions but which only needs to be 12 mm thick.

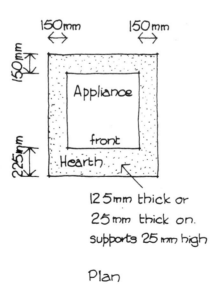

Fig. 12.1 Hearths for back boiler gas appliances

12.2.2 Hearths for oil-fired appliances

If the surface temperature of the floor below the appliance is likely to exceed 100 °C, then a constructional hearth is required. In plan, the hearth should be the same as for a gas appliance, i.e. it should project 150 mm each side and behind the fitting, 225 mm in front and be 125 mm thick. If the hearth is larger than the minimum that this would give, combustible material may be placed on it but no nearer than 150 mm each side and behind the appliance and 225 mm in front.

Oil-fired appliances that are not likely to raise the temperature of the floor surface above 100 °C may stand on a rigid, imperforate sheet of non-combustible material without the need for a constructional hearth.

12.3 Flues and flue pipes

Flue pipes for any oil-fired appliance in which the flue gas temperature is likely to exceed 260 °C may be made in any of the materials listed below, but should only be used to connect the appliance to a chimney. In addition, they should not pass through a roof space.

- Cast iron as described in BS 41: *Specification for cast iron spigot and socket flue or smoke pipes and fittings.*
- Mild steel with a wall thickness of at least 3 mm.
- Stainless steel with a wall thickness of at least 1 mm and as described in BS 149: *Steel plate, sheet and strip.* Part 2: *Specification for stainless and heat resisting steel plate, sheet and strip* for Grade 316 S11, 316 S13, 316 S16, 316 S31, 316 S33, or the equivalent Euronorm 88-71 designation.
- Vitreous enamelled steel complying with BS 6999: *Specification for vitreous enamelled low carbon steel flue pipes, other components and accessories for solid fuel burning appliances with a maximum rated output of 45 kW.*

Flue pipes for all gas-fired appliances and any oil-fired appliances in which the temperature of the flue gas is not likely to exceed 260 °C may be made in any of the following:

- Any of the materials listed above.
- Sheet metal as described in BS 715: *Specification for metal flue pipes, fittings, terminals and accessories for gas-fired appliances with a rated input not exceeding 60 kW.*
- Asbestos cement as described in BS 567: *Specification for asbestos-cement flue pipes and fittings, light quality,* or BS 835: *Specification for asbestos-cement flue pipes and fittings, heavy quality.*
- Cast iron as described in BS 41.
- Any other material fit for its intended purpose.

All flue pipes with spigot and socket joints should be fitted with the socket facing upwards.

A flexible flue liner may be used in a chimney venting a gas appliance or an oil-fired appliance in which the flue gas temperature is not likely to exceed 260 °C, provided that the liner complies with the requirements of BS 715 and the chimney either was built before 1 February 1966, or is already lined or constructed of flue blocks as described in Chapter 11.

12.3.1 Sizes of flue pipes

Flue pipes to balanced flue units, and oil-burning fittings designed for low level flues should be sized in strict accordance with the recommendations (frequently supplied with the appliance) of the manufacturer.

Apart from any special provisions specified in appliances described above, the minimum size for a flue pipe should be not less than the size of the flue outlet on the appliance irrespective of the fuel used.

12.3.2 Direction of flue pipes

Flues should be run vertically wherever possible. Horizontal runs should be avoided except in the cases of a back outlet appliance, when the length of flue between the appliance and a chimney should not exceed 150 mm, or a balanced flue appliance installed as recommended by the manufacturer.

If a bend is required in a flue, the angle it makes with the vertical should not exceed 45°.

12.3.3 Shielding of flue pipes

Flue pipes to gas-fired appliances and oil-fired appliances that have flue gas temperatures less than 260° should be positioned at least 25 mm away from any combustible material. Where they pass through a wall, floor or roof, they should be separated from anything that can burn by a non-combustible sleeve that encloses an air space of not less than 25 mm all round the flue pipe. For a double-walled flue pipe, the 25 mm can be measured from the outside of the inner pipe.

If the wall or floor is of the compartment type as described in Chapters 6 and 8, the flue pipe must be cased with non-combustible material with at least half the fire resistance specified for the wall or floor.

12.3.4 Outlets from flues

The outlet from a flue serving an oil-fired appliance (other than one of the pressure jet or balanced flue type) must be positioned above the roof line in the same manner as that required for a solid fuel appliance and as shown on Figs 11.3 and 11.4 of the previous chapter.

Gas-fired appliances designed to operate with a balanced flue should have their outlet positioned so that there is a free intake of air and dispersal of the products of combustion. The outlet should be at least 300 mm from any opening wholly or partly above the terminal that should be fitted with a guard if it is likely to be touched by a person or damaged in any way. It should also be designed to prevent the entry of anything that might affect the operation of the flue.

The outlet from any other gas-fired appliance must so be sited, at roof level, that air may pass freely across it at all times and be 600 mm from any opening into the building. If the flue, except that to a gas fire, is more than 175 mm across in any direction it should be fitted with a flue terminal.

A balanced flue or a low level discharge flue from an oil-fired appliance can terminate in the same way as a balanced flue from a gas-fired appliance except that it should be not less than 600 mm from any opening into the building.

12.4   Chimneys

The Building Regulations do not deal with the design or construction of the large, free-standing chimneys associated with some buildings. These would, of course, be subject to the skills of a specialist contractor.

The construction of a chimney required to remove the products of combustion from a gas appliance rated up to 60 kW or an oil-fired appliance rated up to 45 kW is principally the same as that serving a solid fuel burning appliance and the whole subject has been dealt with in the previous chapter under Section 11.4.1 to which reference should be made.

12.5   Shielding of appliances

Gas appliances that do not comply with the relevant Parts of BS 5258 or BS 5368 and oil-fired appliances with a surface temperature of the sides or back in excess of 100 °C need to be separated from anything that might burn by a shield of non-combustible material at least 25 mm thick or an air space of not less than 75 mm.

12.6   Air supply

Unless the appliance is room sealed, the room or space in which it is sited should have a ventilation opening of the size required for the type and capacity of the appliance. If this opening is to an adjoining room or space then that too must have a ventilation opening of the same size that vents directly to the open air.

Ventilation openings should not be in fire-resisting walls.

The specific size required for each type of appliance is:

- Gas cooker: An openable window or other means of ventilation of any size.

If the room volume is less than $10\,m^3$ a permanent vent of $5000\,mm^2$ is needed.

- Open-flued gas appliance: $450\,mm^2$ for each kW of input over $7\,kW$.
- Any oil-fired appliance: $550\,mm^2$ for each kW of output over $5\,kW$.

Because of its ventilation requirements, no internal kitchenette (which has to rely solely on mechanical ventilation) should contain an open-flued appliance as there is a risk of flue gas spillage.

12.7 Alternative approaches

The requirements of Regulations J1, J2 and J3 may also be met by compliance with the relevant sections of the following British Standards:

BS 5864: *Specification for installation of gas-fired ducted-air heaters of rated input not exceeding 60 kW (2nd family gases).*

BS 5440: *Installation of flues and ventilation for gas appliances of rated input not exceeding 60 kW (1st, 2nd and 3rd family gases):* Parts 1 and 2.

BS 5871: *Installation of gas fires, convector heaters, fire/back boilers and decorative fuel effect gas appliances:* Parts 1, 2 and 3.

BS 6172: *Specification for installation of domestic gas cooking appliances (2nd family gases).*

BS 6173: *Specification for installation of gas-fired catering appliances for use in all types of catering establishments.*

BS 6798: *Specification for installation of gas-fired hot water boilers of rated input not exceeding 60 kW.*

BS 5410: *Code of practice for oil firing:* Part 1: *Installations up to 44 kW output for space heating and hot water supply purposes.*

12.8 Gas-fired catering appliances

As listed above, the requirements of the Regulations will be met by compliance with BS 6173 where gas-fired catering appliances are to be installed. Much of the design of any catering installation is specialized and unique but here are a few generally applicable points and the following notes are taken from the Standard.

12.8.1 Services

The gas pipework and fittings should be installed in accordance with the appropriate standards. A useful source of information is the British Gas publication IM16: *Guidance notes on the installation of gas pipework, boosters and compressors in customers' premises.*

An emergency isolation system must be fitted. This can comprise either a manually operated valve or an automatic isolating valve system. The manually operated valve is to be fitted near, or just outside, the exit from the kitchen in a readily accessible position. The automatic system should also be operated from a readily accessible position and either comply with BS 5963 or be a low pressure cut-off valve incorporating a small solenoid-operated shut-off valve in the main valve's weep system. Accompanying any automatic cut-off system there must also be an automatic system that proves that all gas supplies to burners and pilot lights have been switched off.

Where manual or automatic cut-off systems can be re-set the following notice must be displayed:

> **All downstream burners and pilot valves must be turned off prior to attempting to restore supply**
>
> **After extended shut-off, purge before restoring gas**

All electrical installations must be in accordance with the Institute of Electrical Engineers' Regulations for Electrical Installation. Each appliance requiring mains should be connected to:

- a fused 13 A socket which can be switched or unswitched, or
- a fused double-pole switch, or
- a heavy duty switched plug and socket.

Water connections should be in accordance with BS 5546, BS 6700 and CP 342: Part 2.

Drainage facilities must be provided both to remove all water used for washing down during general cleaning and for the disposal of water from inside appliances such as boiling pans and steamers. All drains and channels must be accessible for cleaning.

12.8.2   Ventilation

Adequate ventilation must be provided. The BS Standard does not provide any detailed guidance on ventilation requirements as these tend to be of a specialist nature but reference can be made to BS 5720, BS 5925 and the CIBSE publication: *Guide on the ventilation of kitchens.*

Any deep or shallow fat fryers, under-fired grilles and griddles must be provided with canopies venting to the outside air. The canopy must be rigid and stable, extend 200 mm beyond the appliances served and have its bottom edge not less than 2 m from the floor. Where cleanable filters are needed, these must be accessible and the canopy designed so that fat cannot collect and be trapped.

12.8.3   Fire precautions

All appliances and their associated flue systems must be separated from any combustible material in strict accordance with the manufacturer's instructions. This separation should be such as will ensure that the temperature of the combustible material does not exceed 65 °C.

A notice detailing necessary action in the event of a fire in any premises with a deep-fat fryer must be prominently displayed and take the form of:

In the event of fire
close the lid of the fryer

**Do not use water to extinguish the fire
if necessary use the adjacent fire blanket
or a suitable fire extinguisher**

12.9   Unvented hot water systems

The requirements of Regulation G3 do not apply to a hot water storage system that has a storage vessel of 15 litres or less, nor does it apply to a system providing space heating only. Approved Document G3 also points out that the By-laws of the appropriate Water Undertaker apply to these systems.

If it is proposed to employ an unvented system, the local authority will require to know the name, make, model and type of hot water storage system to be installed; the name of the body, if any, that has approved or certified that the system is capable of performing in a way that satisfies the requirements of this Regulation and the name of the body, if any, that has issued any current Registered Operative identity card to the installer or proposed installer of the system.

Any system that incorporated a storage vessel that has no vent pipe must be installed by a competent person, it must have safety devices that prevent the temperature of the stored water from exceeding 100 °C at any time and it must have pipework that safely conveys the discharge of hot water from the built-in safety devices to a point where it is visible but which will not cause any danger to people in the building.

A competent person is defined in Approved Document G3 as one holding a current Registered Operative card for the installation of unvented domestic hot water systems issued by:

- the Construction Industry Training Board (CITB)
- the Institute of Plumbing
- the Association of Installers of Unvented Hot Water Systems

- individuals who are designated Registered Operatives and employed by companies included on the list of Approved Installers published by the British Board of Agrément up to 31 December 1991
- an equivalent body

12.9.1   Unvented systems up to 500 litres storage and 45 kW power input

Any unvented hot water storage system should be in the form of a proprietary package or unit that is:

- approved by a member body of the European Organization for Technical Approvals (EOTA) operating a technical approvals scheme such as the British Board of Agrément
- approved by a certification body accredited by the National Accreditation Council for Certification Bodies (NACCB) which has tested the system against an appropriate standard such as BS 7206: *Specification for unvented hot water storage units and packages*
- the subject of a proven independent assessment that will clearly demonstrate an equivalent level of verification and performance

12.9.2   Direct heating units (500 litres and 45 kW)

Any direct heating unit must have at least two temperature safety devices operating in sequence. One must be a non-self-resetting thermal cut-out in accordance with BS 3955: *Specification for electrical controls for household and similar general purposes,* or BS 4201: *Specification for thermostats for gas burning appliances.* The other must be one or more temperature relief valves to BS 6283: *Safety and control devices for use in hot water systems:* Part 2: *Specification for temperature relief valves for pressures from 1 bar to 10 bar,* or Part 3: *Specification for combined temperature and pressure relief valves for pressures from 1 bar to 10 bar.*

These devices must be in addition to any thermostats that are fitted to maintain the temperature of the stored water.

Alternatively the unit could be fitted with other safety devices that are capable of providing an equivalent degree of safety in preventing the temperature of the stored water from exceeding 100 °C at any time. These devices must be approved by either a member of EOTA or a body possessing NACCB accreditation or be the subject of a proved independent assessment.

A temperature relief valve should be fitted directly onto the storage vessel in such a way as to ensure that the temperature of the stored water does not exceed 100 °C. The valve should be sized so that the discharge rating is not less than equal to the power input to the water. This discharge rating is to be measured in accordance with Appendix F of Part 2 of BS 6283 or Appendix G of Part 3 of the same British Standard.

The discharge from a temperature relief valve must be via a short length of

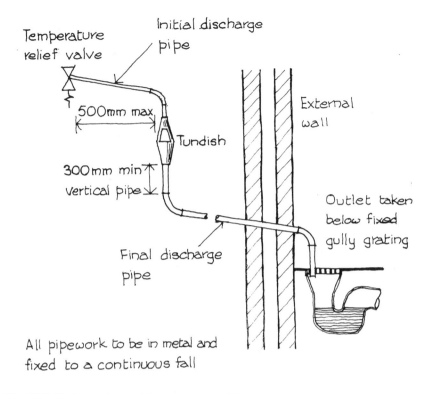

Temperature relief valve

Initial discharge pipe

500mm max

Tundish

300mm min Vertical pipe

External wall

Final discharge pipe

Outlet taken below fixed gully grating

All pipework to be in metal and fixed to a continuous fall

Fig. 12.2 Discharge pipework from an unvented hot water system

metal pipe, of a size not less than the nominal outlet size of the valve, through an air break over a tundish (see Fig. 12.2). Where several valves are fitted, they may discharge through a manifold that must be sized to accept the total discharge from the pipes connected to it.

12.9.3 Indirect heating units (500 litres and 45 kW)

The safety devices listed for direct heating units are also required for indirect units but the non-self-resetting thermal cut-outs should be wired to a motorized valve or a similar suitable device or they should shut off the flow to the primary heater. If the unit incorporates a boiler, the thermal cut-out may be fitted to it.

The temperature relief valve should be sized and fitted and a discharge pipe provided in the same manner as that described for direct heating units.

Indirect heating units that have an alternative direct method of water heating fitted need a non-self-resetting thermal cut-out on this alternative direct source as well as on the indirect.

12.9.4 Unvented systems of more than 500 litres but not more than 45 kW power input

Any systems of this capacity (or the capacity given in the next section) should be individually designed by an appropriately qualified engineer and installed by a competent person as described above.

The system should be fitted with safety devices in accordance with the relevant recommendations of BS 6700: *Specification for design, installation and maintenance of services supplying water for domestic use within buildings and their curtilages.* Non-self-resetting thermal cut-outs appropriate to the heat source should be fitted and discharge pipes installed, all as described above.

12.9.5 Unvented systems of more than 500 litres and more than 45 kW power input

These largest systems should be fitted with the appropriate number of temperature relief valves as recommended in BS 6283: *Safety and control devices for use in hot water systems,* which would give a combined discharge rating at least equal to the power input. Alternatively, the relief valves must be marked with their set temperature in °C and the discharge rating in kW. This marking must be done by a member of the European Organization for Technical Approvals (EOTA), such as the British Board of Agrément, and measured in accordance with either Appendix F of BS 6283: Part 2 or Appendix G of BS 6283: Part 3. The valves must be factory fitted and the sensing element located directly on the storage vessel.

12.9.6 Discharge pipes for all unvented systems

The discharge pipe from the vessel, as far as and including the tundish, is usually a part of the proprietary package. If not, it must be fitted as part of the installation work and, in either case, the tundish must be fixed vertically, in the same space as the water storage system and 500 mm or less from the temperature relief valve.

The pipe from the tundish must be of metal and terminate in a position that will not create a danger to anyone in the vicinity. The minimum diameter of this pipe is one pipe size larger than the nominal outlet size of the temperature relief valve. This assumes a straight pipe of a maximum length of 9 m. If the actual pipe is longer than this or has any elbows or bends, the hydraulic resistance of a 9 m pipe would be exceeded and the diameter would, therefore, need to be increased. Table 12.1 shows the sizing of copper discharge.

The method of sizing is best explained by the following example:

- Check the suitability of a 22 mm copper discharge pipe 7 m long with four elbows serving a G½ temperature relief valve.

Table 12.1   Non-vented heating systems – sizing of final discharge pipe from temperature relief valve

Valve outlet size	Size of initial discharge pipe (mm)	Size of final discharge pipe (mm)	Maximum resistance allowed expressed as a length of straight pipe (m)	Resistance created by an elbow expressed as a length of straight pipe (m)
G½	15	22	up to 9	0.8
		28	up to 18	1.0
		35	up to 27	1.4
G¾	22	28	up to 9	1.0
		35	up to 18	1.4
		42	up to 27	1.7
G1	28	35	up to 9	1.4
		42	up to 18	1.7
		54	up to 27	2.3

(*Source:* Based on Table 1 of Approved Document G3)

- The maximum resistance allowed for a 22 mm copper pipe from a G½ valve, expressed as a length of straight pipe is, 9.0 m.
- The reduction required to take account of the resistance of the four elbows is: 4 × 0.8 m, i.e. 3.2 m.
- The maximum permitted length is: 9.0 − 3.2, i.e. 5.8 m. As this is less than the actual length of 7 m the next larger size of pipe – 28 mm – must be checked and would be found to be satisfactory.

There must be a vertical section of pipe, at least 300 mm long, immediately below the tundish and the rest of the pipework must be fixed with a continuous fall.

It is preferable that the discharge both at the tundish and at the termination of the pipework can be seen but where this is not possible, or presents practical difficulties, one or other of the points of discharge must be clearly visible. Ideally, the final discharge should be to a gully at a point below the grating but above the water seal. Alternatively, it can discharge vertically above an external surface such as a car park, hardstanding or grassed area at a height not exceeding 100 mm above the surface. If there is a possibility that children might come into contact with the end of the discharge pipe, a wire cage or similar guard must be fitted that will prevent injury but not restrict the view of the pipe.

High level discharge is also permitted but it must still be visible and if a hopper head and downpipe are provided to receive it, they must be of metal. If the discharge is onto a roof, it must be capable of withstanding the high temperature water, the discharge must be at least 3 m from any plastic guttering that would collect the discharged water and the tundish must also be visible. Note, in this connection, that as the discharge would consist of scalding water or steam, asphalt roofing and bituminous roofing felt would be damaged by it.

A single pipe can serve a number of discharges but the maximum should be restricted to six so that any installation discharging can be traced with reasonable ease. The common discharge pipe must be sized in relation to the largest individual discharge pipe connected to it.

If unvented hot water storage systems are installed in premises where the safety devices may not be apparent to the occupants – dwellings such as those occupied by blind, infirm or disabled people – consideration should be given to fitting an electronically operated device to warn when a discharge takes place.

12.9.7   Electrical connections

Electrical non-self-resetting thermal cut-outs should be connected to either the direct heat source or the indirect primary flow control device in strict accordance with the current Regulations for Electrical Installations of the Institute of Electrical Engineers.

12.10	Heating controls and insulation of ducts, hot water cylinders and pipes

In addition to the requirements about the conservation of fuel and power by the reduction of heat losses through the external fabric of the building, the Regulations also require that provision shall be made to save energy by reducing any waste that could occur in the space heating and hot water systems of the building.

12.10.1   Space heating controls

Both temperature controls and time controls are required to satisfy the requirements of the Regulation. Acceptable temperature control can be achieved by the provision of thermostatic radiator valves, thermostats or any other equivalent form of temperature-sensing device, the latter two to be fitted to each part of the system designed to be separately controlled. In addition, if the system uses hot water as the transfer medium, an external temperature sensor and a weather compensator controlling the flow of the water should be installed.

Suitable time controls depend on the output of the system. If the output is

100 kW or less, a simple clock control that is manually set to provide heat only when the building is occupied would meet the requirements. A heating system with an output in excess of 100 kW requires optimizing controllers that set the start times to correspond to the rate at which the building will cool down and then heat up again with the on–off heating cycle.

Additional controls may be fitted to prevent any frost or condensation damage during the period when the normal heating service would be switched off.

In large heating installations, it is quite common to find a number of boilers installed that are switched on and off in sequence as the heating demand rises and falls. Approved Document L1 states that such a multiple boiler installation is to be fitted with a sequence controller that would detect the variations in heating demand and start or stop the modulate boilers in combinations that are effective for the purposes of conservation of energy.

An alternative would be to follow the recommendations of BS 6880: *Code of practice for low temperature hot water heating systems of output greater than 45 kW*, or CIBSE Application Manual AM1: *Automatic controls and their implications for system design.*

12.10.2   Control and insulation of hot water storage vessels

The requirements of the Regulation will be met if the storage system has a heat exchanger with sufficient heating capacity for effective control and is fitted with a thermostat and timer.

The adequacy of the heat exchanger can be assured by a compliance with either BS 1566: *Copper indirect cylinders for domestic purposes* or BS 3198: *Specifications for copper hot water combination units for domestic purposes.*

The thermostat should shut off the supply of heat when the set water temperature has been reached and it should also be interconnected to the space heating thermostats so as to switch off the boiler when no heat is required.

The timer can be either part of the central heating control system or a local device, in either case, it should shut off the supply of heat during any periods when hot water is not required.

Hot water vessels can comply with BS 1566, BS 3198 or BS 853: *Specification for calorifiers and storage vessels for central heating and hot water supply.*

Vessels complying with either of the first two Standards should have factory applied insulation that restricts the heat loss to 1 W/litre or less. This can be achieved by a 35 mm coating of PU-foam with a minimum density of 30 kg/m and a zero ozone depletion potential. Vessels to BS 853 should be provided with a 50 mm layer of an insulating material with a thermal conductivity of 0.045 W/m K, or other material applied in a thickness to give an equivalent performance. Many manufacturers also provide a metal outer casing to provide physical protection of the insulation.

12.10.3   Insulation of pipes and ducts

In some installations, the heat loss from a pipe or, more rarely, from the surface of a duct is allowed in the calculation of heating surfaces required. If not, then the pipe or duct should be insulated to reduce energy wastage.

Pipe insulation generally should consist of:

- a material with a thermal conductivity not greater than 0.045 W/m K (which covers most of the commonly available insulating products), and
- a thickness equal to the outside diameter of the pipe up to a maximum of 40 mm.

The primary flow and return pipes and the vent pipe connected to a hot water storage vessel complying with BS 1566 or BS 3198 should all be insulated with a 15 mm thick sleeve of an insulating material with a thermal conductivity of 0.045 W/m K for a distance of 1 m from the vessels or up to the point where the pipes are concealed. The thickness of the insulation on pipes connected to a vessel complying with BS 853 should be equal to the outside diameter of the pipe up to a maximum of 40 mm.

Alternatively, Approved Document L1 states that pipe and duct insulation should be provided in accordance with the recommendations of BS 5422: *Method for specifying thermal insulating materials on pipes, ductwork and equipment.*

The British Standard sets out economic thicknesses for both duct and pipe insulation, taking into account the average cost of the materials and the labour cost of installation against the cost of the heat lost over a 5-year period. The Standard gives values for a range of thermal conductivities of insulating material from 0.02 to 0.07 W/m K for ductwork and from 0.025 to 0.045 W/m K for pipes. The following notes and Table 12.2 are based on one, average, value. Guidance should be sought from BS 5422 if materials with other values are to be used.

The economic thickness for insulation on ductwork carrying warm air depends on the temperature difference between the air inside and outside

Table 12.2   Minimum thickness of pipe insulation with material of a thermal conductivity of 0.035 W/m K

Pipe size (mm)	Minimum thickness of pipe insulation (mm) for:					
	Central heating		Hot water		Cold water	
	Heated areas	Unheated areas	Heated areas	Unheated areas	Indoors	Outdoors
15	21	32	14	17	62	279
22	30	34	16	20	20	47

(*Source:* Based on BS 5422: 1990)

the duct and is as follows, based on material with a thermal conductivity of 0.04 W/m K:

Temperature difference (K)	Thickness of insulation (mm)
10	38
25	50
50	63

The economic thickness for pipe insulation for heating and hot water systems is shown in Table 12.2 based on insulation with a thermal conductivity of 0.035 W/m K.

As can be seen, different thicknesses apply depending on which system is being insulated and whether the pipe is running through a heated or unheated area.

Table 12.2 also shows the same insulating material thicknesses for cold water pipes necessary to give protection against freezing. The Standard points out that if the ambient temperature remains low for a long time and there is no movement of water along the pipe, the use of insulation alone will not afford a complete protection against internal freezing. For the smaller pipes it is not practicable to install thermal insulation of sufficient thickness to avoid entirely the possibility of ice formation overnight in sub-zero temperatures. The problem is more acute the smaller the pipe, due to the lesser heat capacity, and, as can be seen from the table, the thickness required to afford reasonable protection to a 15 mm pipe in an outside location is a totally impracticable 279 mm. Changing the pipe size to 22 mm reduces this to a realistic 47 mm and is a recommended practice, unless some form of suitable pipe heating is employed.

The need to increase the thickness of insulation in unheated areas is also referred to in the BRE Report BR 262: *Thermal insulation: avoiding risks.*

12.11   Lighting

The Regulation requirement states that lighting systems must be designed and constructed to use no more fuel and power than is reasonable in the circumstances. However, this requirement does not apply where the area of floor to be artificially lit is less than $100\,m^2$.

The method adopted in Approved Document L1 to achieve compliance with this Regulation is to define standards for the efficacy of lamps used and suitable lighting controls that will avoid any unnecessary lights being on during the times when rooms or spaces are unoccupied. Because of their specialized nature, none of the requirements of Approved Document L1 applies to emergency escape lighting, which is dealt with in Chapter 15.

12.11.1   Minimum efficacy of lamps

To meet the requirements of the Regulation, the lamps used in any of the

Table 12.3 Types of high efficacy lamps

Light source	Types of lamp
High-pressure sodium Metal halide Induction lighting	All types and ratings
Tubular fluorescent	All 25 mm diameter (TB) lamps provided with low-loss or high-frequency control gear
Compact fluorescent	All ratings over 11 W

(*Source*: Based on Table 9 of Approved Document L1 (1995 edition))

types of building covered in this book must either have an initial (100 hour) efficacy of not less than 50 lumens per circuit watt or at least 95 per cent of the lamps must be one or more of the types shown in Table 12.3.

12.11.2   Lighting control

Lights in buildings for assembly or residential use should be fitted with local switches, time switches or photo-electric switches as appropriate.

Local switches include:

- manual switches such as rocker switches, press buttons and pull cords
- remote control switches operated by infra-red transmitter, sonic, ultra-sonic or telephone handset controllers
- automatic switching systems, including controls that switch off the lighting when they sense the absence of occupants.

These local switches should be located within each working area or at the boundary between a working area and a general circulation area. Manually operated switches must be readily accessible and the maximum distance on plan from a local switch to the furthest lighting fitting should generally not exceed either 8 m or three times the height of the light fitting above the floor.

Automatically operated switching systems should be designed to operate in a way that avoids endangering the passage of people.

The guidance given in the CIBSE publication: *Code for interior lighting* can be followed as an alternative to the details given above, but the alternative designs based on this Code should perform at least as well as the minimum standards laid down in Approved Document L1.

Chapter 13

WASHROOMS, TOILETS AND ABOVE GROUND DRAINAGE

| 13.1   The Building Regulations applicable |

All buildings require sanitary facilities and the building uses covered in this book are no exception. The Building Regulations relating to these installations and, therefore, reviewed in this chapter are:

B3   Internal fire spread (structure)
F1   Means of ventilation
G1   Sanitary conveniences and washing facilities
H1   Sanitary pipework and drainage
M3   Access and facilities for disabled people

Regulation B3 deals with the fire risk posed where pipes penetrate compartment walls and floors; F1 requires that there shall be adequate ventilation for the people in the building; G1 states that there shall be adequate sanitary conveniences (by this is meant water and chemical closets) and wash basins, with hot and cold water, in rooms separated from places where food is prepared; H1 is concerned that any system carrying foul water from appliances shall be adequate; and M3 requires that if sanitary conveniences are provided in the building, reasonable provision shall be made for disabled people. The way the requirements of Regulation M3 are applied to new buildings, extensions to buildings and alterations to buildings is set out in detail in Chapter 16.

Additional guidance on sanitary provisions and access is given in Design Note 18 published by the Department for Education with the title of *Access for disabled people to educational buildings*. Where they are relevant to the provision of sanitary facilities, the recommendations are noted in this chapter and, where relevant to access, in Chapter 16.

In this chapter the more commonly used name of 'toilet' is employed for what is described in the Regulations as Sanitary Accommodation and water closet or WC is used for the fitting, although the Regulations do allow the use of chemical closets in buildings (subject to the same rules) where there is no suitable water supply or means of disposal of foul water.

13.2 Sanitary provisions generally

Approved Document G1 states that the requirements of Regulation G1 will be met by a sufficient number of sanitary conveniences of the appropriate type for the sex and age of the people using the building. Approved Document G1 does not specify the precise standards of provision in buildings for public assembly and residential use but refers to relevant Acts and British Standards, particularly BS 6465: Part 1: *Code of practice for scale of provision, selection and installation of sanitary appliances.* The BS does not cover all buildings, notably health care premises and sports facilities (except for swimming pools). The scale and installation of sanitary equipment in hospitals and other health care buildings can be found by reference to the appropriate *Health Building Note,* published by NHS Estates, covering the particular application, or to *Activity Data Base,* also by NHS Estates. This data base is a computerized summary of all the published recommendations for health care buildings.

13.2.1   Scale of provision

The numbers and types of fittings required in buildings for public entertainment, housing for the elderly, hotels, educational establishments and schools are shown in Table 13.1. Where relevant, it has been assumed in setting the standards shown in this table that the mix of users will be 50 per cent male and 50 per cent female. Where a different proportion is to be expected the accommodation levels should be adjusted.

A separate unisex provision should be made for the purpose of changing nappies. This should contain WC and basin provisions that are suitable for children, appropriate disposal facilities and suitable materials either to wipe or to cover the surfaces used.

The provisions to be made for staff toilets where appropriate is shown in Table 13.2. This follows the recommendations in BS 6465. If these sanitary facilities are also to be used by the public, the number of conveniences shown in Table 13.1 should be increased by at least one for each sex.

In certain situations, security may require separate facilities to be provided for members of the public; these should be located in or adjacent to the public area.

13.2.2   Design recommendations

The design of a WC, a urinal and a wash basin should provide a surface that is smooth, non-absorbent and easily cleaned. The appliance may be floor mounted or wall hung, the latter arrangement being recommended in any premises where regular floor cleaning is necessary. All junctions between the appliances and the floor or wall should be sealed.

The proper and thorough cleaning of toilets is important and BS 6465

Table 13.1 Standards of sanitary provision

Accommodation	WC	Urinal	Wash basin	Bath or shower	Comments
Buildings for public entertainment					
Without licensed bar					
for females	2 (up to 40) 3 (41 to 70) 4 (71 to 100) plus 1 per 40 over		1, plus 1 per 2 WCs		
for males	1 (up to 250) plus 1 per 500 over	2 (up to 100) plus 1 per 80 over	1 per WC plus 1 per 5 urinals		
With licensed bar (as above but plus the following)					
for females	1 (up to 12) 1 (13 to 30) plus 1 per 25 over		1 per 2 WCs		Accommodation based on equal proportions of male and female customers at 4 persons per 3 m² effective drinking area in the bar
for males	1 (up to 150) plus 1 per 150 over	2 (up to 75) plus 1 per 75 over	1 per WC plus 1 per 5 urinals		
Housing for the elderly					
Sheltered housing					
per dwelling	1		1	1	Possibly an additional WC in the bathroom
Grouped accommodation					
per dwelling	1		1	1 per 4 apartments	Sitz bath or level access shower
Common room	1				Available for visitors
Residential/nursing homes					
for residents	1 per 4 persons		1 per room	1 per 10 persons	Possibly Sitz bath or level access shower
for staff	at least 2		1		Wash basin in WC
for visitors	1		1		Wash basin in WC
Hotels					
With en-suite rooms					
per guest bedroom	1		1	1	In en-suite bathroom
per staff bedroom	1 per 9		1 per 9	1 per 9	In staff bathroom
Without en-suite rooms					
per guest	2 per 9		1 per 9	1 per 9	Bath/shower, basin and one WC in a shared bathroom
per bedroom			1		
per guest in a dormitory			1 per 9		

(continued)

Table 13.1   (continued)

Accommodation	WC	Urinal	Wash basin	Bath or shower	Comments
Educational establishments					
Special schools					
Boys	1 fitting per 10 pupils, not more than ⅔rd to be urinals		1 in each washroom, 2 per 3 fittings	Sufficient showers should be provided for physical education	'Fitting' means either a WC or a urinal
Girls	1 per 10		1 in each washroom, 2 per 3 WCs		
Primary schools					
Boys	1 fitting per 10 pupils under age 5 and 1 per 20 over age 5, not more than ⅔rd to be urinals		1 in each washroom, 2 per 3 fittings	Sufficient showers should be provided for physical education	'Fitting' means either a WC or a urinal
Girls	1 per 10 pupils under age 5 and 1 per 20 over age 5		1 in each washroom, 2 per 3 WCs		
Secondary schools					
Boys	1 fitting per 20 pupils, min 4 fittings, not more than ⅔rd to be urinals		1 in each washroom, 2 per 3 fittings	Sufficient showers should be provided for physical education	'Fitting' means either a WC or a urinal
Girls	1 per 20 pupils, min 4 WCs		1 in each washroom, 2 per 3 WCs		Provision may be required for the disposal of sanitary dressings
Nursery and play schools					
All pupils	1 per 10, min 4		1 per WC		Additionally, 1 sink per 40 pupils should be provided
Boarding schools					
All pupils	1 per 5 boarding pupils		1 per 3 up to 60 plus 1 per 4 for next 40 and 1 per 5 over 100	1 per 10 boarding pupils. At least 25% must be baths	Where day pupils' sanitary accommodation is also available for boarders, these requirements may be reduced as approved by Dept for Education
Swimming pools					
(Assuming equal proportions male and female swimmers)					
Male	2; up to 100 plus 1 per 100 over	1 per 20	1 per WC plus 1 per 5 urinals	1 per 10	Toilets should be provided for spectators as given above for buildings for public entertainment
Females	1 per 5 up to 50 plus 1 per 10 over		1, plus 1 per each 2 WCs	1 per 10	

(*Source*: Based on BS 6465: Part 1: *Code of practice for scale of provision, selection and installation of sanitary appliances*)

Note: In addition to the provisions given above, sanitary facilities should be provided for staff in accordance with Table 13.2.

Table 13.2   Scale of provision for staff toilets

Any group of staff			Alternative provisions where the staff is male only		
Number of persons at work	Number of WCs	Number of wash basins	Number of men at work	Number of WCs	Number of urinals
I to 5	1	1	I to 15	1	1
6 to 25	2	2	16 to 30	2	1
26 to 50	3	3		2	2
51 to 75	4	4	46 to 60	3	2
76 to 100	5	5	61 to 75	3	3
Over 100	One WC and one wash basin for each additional 25 persons (or fraction of 25)		76 to 90	4	3
			91 to 100	4	4
			Over 100	One WC and one wash basin for every additional 50 men (or fraction of 50)	

(*Source:* Based on Table 4 of BS 6465: Part 1: *Scale of provision, selection and installation of sanitary appliance*)

Note: Sanitary facilities should be provided for staff as indicated by this table in addition to those shown in all the building types listed in Table 13.1.

recommends that storage space for cleaning materials and implements should be provided along with a supply of hot and cold running water with a suitable sink, all in the vicinity of any non-domestic installations. It is also essential that shower trays and their surrounding areas can be easily cleaned to minimize the risk of cross-infection of foot ailments, particularly in the case of communal showers.

Any flushing apparatus should be capable of cleansing the receptacle effectively and no part of the receptacle should be connected to any pipe other than a flush pipe or a soil pipe branch.

The wall and floor adjacent to a bowl or trough urinal should be impervious and the appliance should be mounted at 600 mm from the floor to the front lip but provision should be made for children and persons of lesser stature.

A WC should discharge through a trap and branch pipe to a soil pipe or drain, a urinal can discharge through a grating, trap or branch pipe to a soil pipe or drain and a wash basin should discharge through a trap and pipe connected to a soil pipe. If the wash basin is located on the ground floor it can discharge through a trap and into a gully or directly into a drain.

Separate toilet facilities should be provided for males and females unless each WC compartment is intended to be used by one person at a time, is a separate room and has a door that can be secured from inside.

Any WC compartments should be fully self-contained (including hooks to avoid having to put clothes on the floor) and if the compartments are in a range of cubicles, the partitions should finish between 100 and 150 mm from the floor and be not less than 2 m high. Male and female compartments should both contain the means to dispose of soiled contingency aids and female WC compartments should also provide for the disposal of soiled sanitary dressings by means of an incinerator, a macerator or sealed bags or bins. There should also be suitable material provided to wipe or cover the WC seats before use by a member of the public.

Wall and floor surfaces should be smooth, floors should be impervious, non-slip and coved to the walls. Any removable panels concealing pipework should be locked in position so as to be incapable of removal except by an authorized person. Anti-graffiti measures should also be included in the design.

Access to a WC compartment should be separated by a door from any workplace or room where food is prepared or washing-up done, nor should the access be from a bedroom or dressing room unless there is alternative WC accommodation.

The door to the compartment should not interfere with any circulation spaces or stairs and be fitted with a simple lock that is easily operated by the user and can be readily released from outside for the purpose of access in the case of an emergency. This easy release is particularly important where the use is by the elderly, by disabled persons or by children.

There should not be any steps in a toilet.

Adequate ventilation is necessary, as described in Chapter 9, consisting of windows or skylights having a total area of 1/20th of the floor area. Where the WC compartments do not reach the floor, the whole area can be treated as one for the purpose of calculation of ventilation area. Alternatively, mechanical ventilation can be provided, capable of producing three air changes per hour. The fans may be operated intermittently, usually through the light switch, and should be fitted with a 15 minute overrun. Extract fans are required where baths or showers are installed to reduce the possibility of condensation.

The lay-out of the appliances within the toilet including the peripheral fittings such as hand dryers, towel cabinets, disposal bins, sanitary towel and condom dispensers and mirrors should be considered and arranged to allow adequate circulation area. Hand-washing appliances and drying equipment should be located between the WCs or the urinals and the exit but positioned so as to avoid creating any congestion.

A drinking water fountain should be provided within the vicinity of, but not inside, WC cubicles. If it is in the toilet area it should be sited as far as possible from the WCs and urinals and should be of the shrouded nozzle type discharging above the spill-over level of the bowl.

BS 6465 makes certain specific recommendations related to the building use:

- In a theatre or any other with a transitory population, the location of toilet facilities should be well signed and, where there is more than one, directions to the nearest alternative location should be signed.
- Where the sanitary facilities are provided for use by elderly persons, WC compartment and bathroom doors should be capable of being opened or removed from the outside and hand grips should be installed beside any WC, bidet, bath or shower. In grouped, one-person apartments, it is usual to provide shared bathrooms. If there is a group of not less than four bathrooms, one shower compartment may be provided instead of one of the bathrooms. A wall-mounted seat should be provided within the shower cubicle.
- Where schools are also to be used for adult education, the choice of fittings, the layout and the location of toilets should be appropriate to both adult and child usage. Showers should be provided for the use of children engaged in physical education. Medical inspection rooms are required and should be fitted with a sink and a separate drinking water supply. A WC and wash basin should be provided in a compartment adjacent to the medical room.
- The provisions shown in Table 13.1 for buildings for public entertainment are the minimum needed; where the premises are subject to an entertainment licence, the scale of provision and the location and arrangement of facilities will be under the direction of the licensing authority. Premises that have distinct intervals in their programme, such as a theatre or a single-screen cinema, require a higher standard of provision than those, like a library or multiple-screen cinema, that have a continuously changing population.
- In hotels, a clearly marked drinking water supply should be provided in each bathroom in addition to the appliances shown in Table 13.1. Any toilets provided for non-residents should be sited so as to avoid excessive travel distances and equipped with washing facilities adjacent to the WCs and urinals.
- Toilets for bathers in a swimming pool should be between the changing area and the pool. They will usually precede, or be combined with the per-cleanse area. If unisex changing facilities are provided, the showering facilities can also be unisex. Toilets provided for spectators should be in addition to and separate from those provided for the bathers.

13.2.3 Washing facilities

Suitable and sufficient washing facilities, including showers if required by the nature of the building use, are to be provided at easily accessible places. They should also be in the immediate vicinity of every sanitary convenience, whether or not they are provided elsewhere and in the vicinity of any changing rooms required.

Wash basins and drying facilities should be in the same room as the room containing the WCs or urinals and, as already mentioned, between them and the exit, or in an adjacent space, sited, designed and installed so as not to be prejudicial to health. The wash basin must have a supply of hot water that may be from a central source or from a unit water heater as well as a piped supply of cold water, be provided with soap or suitable detergent and hand drying facilities, which may be warm air, a hand towel, a roller towel or paper towels.

Further provisions are required in sanitary accommodation associated with food rooms, see 13.2.4 and in toilets for the disabled, see 13.2.5 below.

13.2.4   Sanitary accommodation in association with food rooms

The Food Hygiene (General) Regulations, made under the Food Act of 1990, lay down standards to be observed when sanitary accommodation is provided in a building where food is handled or prepared. The uses to which these Regulations apply include food shops as well as kitchen, canteens and similar spaces.

In these areas the sanitary accommodation should:

- be so placed that no offensive odours can get into any food room (as already mentioned the requirements of Regulation G1 demand that any space containing a WC or urinal is to be separated by a lobby from any room where food is prepared or washing-up done)
- not communicate directly with a food room to be used for handling open food, i.e. food that is not in a sealed container, which would exclude the risk of any contamination
- be suitably and sufficiently lit and ventilated
- not be used as a food room
- have a notice displayed near every sanitary convenience requiring users to wash their hands after using
- be kept clean and in efficient order

Suitable and sufficient wash basins are to be provided in positions that are conveniently accessible each with a supply of hot and cold water (if no open food is being handled a cold water supply only is sufficient), soap, or a suitable detergent, a nail brush and a towel or suitable means of hand drying. The basins must not be used for any purpose other than that of securing personal cleanliness.

There is to be a constant, clean and wholesome supply of water to the food room and no cistern supplying water to a food room shall also supply a sanitary convenience otherwise than through another efficient cistern or other flushing apparatus suitable for the prevention of contamination of the water supply.

13.2.5 Toilets for the disabled

Regulation M3 is much more specific on the subject of facilities for anyone with disabilities and requires that, where sanitary conveniences are provided in a building, reasonable provision shall be made for disabled people.

The Regulations recognize that there are two levels of difficulty to be accommodated. Those experienced by wheelchair-borne persons and those met by anyone who has a disability but can walk, albeit they may be unsteady on their feet or may require some support to sit down or stand up.

Different considerations apply to sanitary accommodation for employees to that provided for visitors or customers. Employees may not need as much assistance as visitors, because of familiarity with the building; furthermore, such assistance as is needed by an employee is likely to be provided by a person of the same sex whereas the visitor or customer would probably be accompanied by someone of the opposite sex.

The number and location of WCs for the disabled depend on the size of the building and the ease of access to the toilet. Wheelchair users should not have to travel more than one storey to reach a suitable WC. It should also be borne in mind that some disabled people need to get to a WC quickly, so travel distances must be kept as short as possible.

Two forms of accommodation are defined in Approved Document M3: 'unisex' and 'integral'. Unisex facilities are available and suitable for either men or women and are approached separately from other sanitary accommodation. The advantage of this arrangement is that it is more easily seen, it is more likely to be available when needed, it permits assistance by a person of the opposite sex and, overall, it tends to be more economical of space than an integral arrangement.

Integral facilities are specially designed WC compartments provided within general sanitary accommodation. This arrangement requires a duplication of the disabled provision but, where two or more are required by reason of the number of people in the building, it may be the best answer.

Approved Document M3 states that wheelchair users' needs will be fulfilled by:

- the provision of unisex accommodation for visitors or customers
- the provision of either unisex or integral accommodation for staff
- a WC compartment for the disabled on alternate floors provided that the travel distance from a workstation to the WC does not exceed 40 m (if the building has stair access only, suitable sanitary accommodation for wheelchair users should be provided in the principal entrance storey unless that storey contains only the main entrance and the stairs)

In an educational establishment, the minimum provision is one WC compartment accessible to wheelchair users in addition to the minimum requirements set down in the Education (School Premises) Regulations.

In nursery schools, the normal sanitary provision, which would include facilities for bathing and changing a child in private, should usually be suitable for disabled children but a suitable WC compartment should be provided for adult users.

In infant and junior schools, the provision for physically disabled children must include the means for bathing and changing in private. There must also be provisions for adult disabled persons.

The WC compartments should:

- be similar in layout
- have enough space for necessary wheelchair manoeuvres
- allow for frontal, lateral, diagonal and backward transfer onto the WC
- have facilities for hand washing and drying within reach from the WC, prior to transfer back to the wheelchair
- have space to allow a helper to assist in the transfer

If there is more than one WC compartment for wheelchair users, the opportunity should be taken to provide both left- and right-hand transfer layouts. A suitable layout and dimensions are shown on Fig. 13.1.

The needs of ambulant disabled people would be met by the provision of an adequately sized WC compartment and the provision of support rails, all as shown in Fig. 13.2. At least one such specially designed compartment should be provided in each range of WC compartments included in storeys that are not designed to be accessible to wheelchair users.

In a hotel, the requirements of Regulation M can be met by the provision of suitable en-suite accommodation in those guest bedrooms that are designed to be accessible by disabled people. Alternatively, if the general arrangement of the hotel is not of the en-suite type, unisex accommodation for disabled people should be provided near to the bedrooms. This is dealt with in more detail in Chapter 16.

13.2.6 Drinking facilities

A clearly labelled supply of drinking water is to be provided in rooms intended for food preparation or beverage making.

Where there are no specific beverage-making facilities, adequate supplies of drinking water are to be supplied, readily accessible and clearly marked. In addition there is to be a supply of cups or other drinking vessels unless the supply is in the form of a fountain from which a person can drink easily. These drinking water points may be in a toilet area but not inside a WC cubicle. If they are in a toilet area they should be remote from the WCs and urinals and be of the type with a shrouded nozzle that discharges above the spill-over level of the bowl.

13.2.7 First Aid rooms

All First Aid rooms should be supplied with a sink or basin fitted with taps that

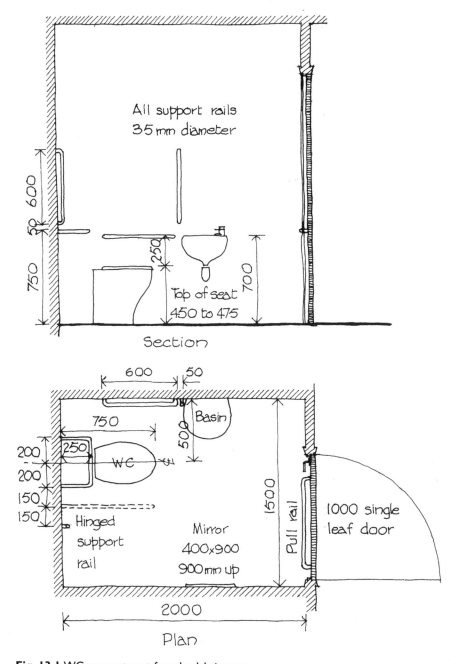

All support rails
35 mm diameter

600

50

750

250

700

Top of seat
450 to 475

Section

600

50

750

250

500

WC

Basin

200

200

150

150

Hinged
support
rail

1500

Pull rail

1000 single
leaf door

Mirror
400x900
900mm up

2000

Plan

Fig. 13.1 WC compartment for wheelchair users

251

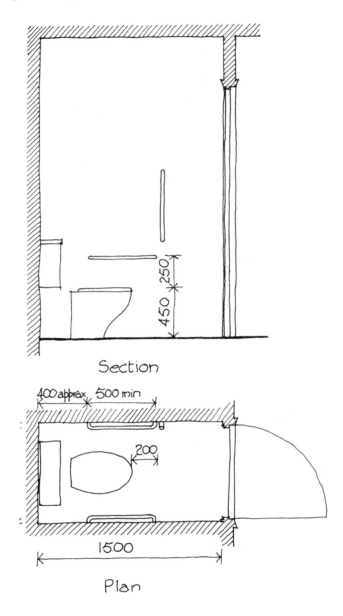

Fig. 13.2 WC compartment for people with limited mobility

can be operated by the wrist or elbow or by remote control. There should also be a supply of drinking water as described in the last section.

In addition, all First Aid rooms should have a WC and washing facilities, as described above, provided in an adjacent compartment.

13.2.8   Cleaners' rooms

Cleaners' rooms are necessary on each floor in a large building. Where there will be fewer than 5 cleaners per floor (assuming approx. 400 m² per cleaner) the provision can be a cleaners' room on every other floor. The room should contain a cleaners' sink or bucket sink, a wash basin and drying facilities plus adequate storage for cleaning appliances and materials.

13.3   Sanitary pipework

Pipe sizes given in Approved Document H1 and in this section are nominal sizes in round figures approximating to the manufactured sizes. Equivalent pipe sizes for individual pipe standards can be found from the manufacturer or in BS 5572.

The capacity of the pipework system should be large enough to carry the expected flow at any point. This capacity depends on the size and gradient of the pipes and the flow depends on the type, number and grouping of the appliances. The minimum pipe sizes in common use are capable of carrying the flow from quite large numbers of appliances.

As appliances are seldom used simultaneously, the flow rate to be accommodated is not the total of their respective discharges but a reduced figure that takes into account the diversity of use (see Chapter 14, Section 14.3).

The requirements of Regulation H1 will be met if the sanitary pipework:

- conveys the foul water to a foul water outfall
- minimizes the risk of blockage
- prevents foul air generated by the drainage system from entering the building
- is ventilated
- is accessible for clearing any blockages

13.3.1   Materials for pipework

Approved Document H1 lists a number of British Standards in which materials are specified that are suitable for use in sanitary pipework. These are shown in Table 13.3.

If different metals are used they should be separated by non-metallic material to prevent electrolytic action, and all pipes should be firmly supported without restricting thermal movement.

13.3.2   Airtightness

All pipes, fittings and joints should be capable of withstanding an air or smoke test with a positive pressure of at least 38 mm water gauge for at least 3

Table 13.3 British Standards specifying materials for pipework

Use	Material	British Standard
Pipes	Cast iron	BS 416 and BS 6087
	Copper	BS 864 and BS 2871
	Galvanized steel	BS 3868
	uPVC	BS 4514
	Polypropylene	BS 5254
	Plastics: ABS, MUPVC, polyethylene, polypropylene	BS 5255
Traps	Copper	BS 1184
	Plastics	BS 3943

(*Source*: Based on Table 5 of Approved Document H1)

Note: Some of these materials may not be suitable for conveying non-domestic effluent.

minutes. During this time, every trap should maintain a water seal of at least 25 mm. Smoke testing is not recommended for uPVC pipework.

13.3.3 Traps

A trap or water seal should be fitted at all points of discharge into the drainage system to prevent foul air from entering the building. Under working conditions and under test a trap should retain a minimum seal of 25 mm. Minimum trap sizes and depths of seal are given in Table 13.4.

All traps should be accessible for the purpose of clearing blockages. If the trap forms part of a fitting, the whole appliance must be removable, otherwise the trap should be fitted directly after the appliance and should be either removable or fitted with a cleaning eye.

13.3.4 Branch pipes

Branch pipes from appliances on upper floors should discharge into another branch pipe or a soil stack. If the appliance is on the ground floor the branch pipe may be connected to a soil stack, a stub stack, or directly to a drain. If the pipe carries only waste water it may, alternatively, discharge into a gully, terminating between the grating or sealing plate and the surface of the water seal in the trap.

Table 13.4 Minimum trap sizes and depth of seal

Appliance	Diameter of trap (mm)	Depth of seal where the discharge is to a soil pipe (mm)	Depth of seal where the discharge is to a gully (mm)
Wash basin	32	75	75
Bidet	32	75	75
Sink	40	75	38
Bath	40	75	38
Shower	40	75	38
Urinal stall (1 to 6 persons)	65	50	75
Urinal bowl	40	75	75
WC pan	75 (siphonic only)	50	Not applicable
Food waste disposal unit	50	75	75
Sanitary towel macerator	40	75	75

(*Source*: Based on Table 2 and Table A2 of Approved Document H1)

The branch connections to a stack pipe should be offset to avoid causing crossflow into other branch pipes and, in buildings up to five storeys, they should not discharge into a stack lower than 750 mm above the invert of the tail of the bend at the foot of the soil stack, as shown in Fig. 13.3. Over five storeys and up to 20 storeys the ground floor appliances should discharge into their own stack and in buildings over 20 storeys, both the ground and first floor appliances should discharge into their own soil stack.

A branch pipe from a WC should only discharge directly into a drain if the drop from the crown of the trap to the invert of the drain is less than 1.5 m.

The size of a branch pipe serving a single fitting should be not less than the diameter of the trap to which it is connected. If it serves more than one appliance, the minimum size for up to eight WCs should be 100 mm with a fall of between 9 and 90 mm/m, and for up to four wash basins it should be 50 mm with a fall of between 18 and 45 mm/m. The length of the branch serving the eight WCs should not exceed 15 m and that to the four wash basins should not be greater than 4 m with no bends.

Bends in branch pipes should be avoided as far as possible. Where bends have to be fitted, they should be of as large a radius as practicable and not less than 75 mm (centreline radius) in pipework of 65 mm diameter and under.

Junctions should be formed with a sweep of 25 mm radius if the pipe is under 75 mm diameter and 50 mm radius if the pipe is 75 mm or more, or, in either case, connecting at 45°.

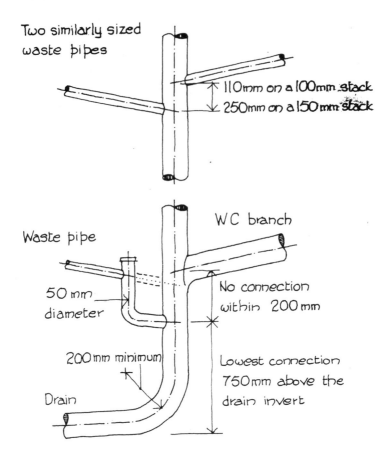

Two similarly sized
waste pipes

110mm on a 100mm stack
250mm on a 150mm stack

WC branch

Waste pipe

50 mm
diameter

No connection
within 200mm

200mm minimum

Lowest connection
750mm above the
drain invert

Drain

Fig. 13.3 Connection of branch pipes to soil pipe

13.3.5 Ventilation of branch pipes

Branch pipes can be unvented up to the lengths shown in Fig. 13.4; beyond these limits, a branch ventilating pipe must be fitted to prevent pressures developing which could cause the water seal in the traps to be lost.

The ventilating pipe can be taken to external air or to a soil stack to form a modified single stack system or to a ventilating stack to form a ventilated system. The latter is only likely to be the preferred arrangement where there are a large number of ventilated branches or the ventilating pipe runs to the soil stack are very long. It must be connected to the branch pipe within 300 mm of the trap and, if it is run into the soil stack, connected above the 'spill-over' level of the highest appliance served (see Fig. 13.5). Branch ventilating pipes run to the external air should terminate in the same manner as the ventilating part of a soil stack, i.e. at least 900 mm above any opening into the building within 3 m of the pipe.

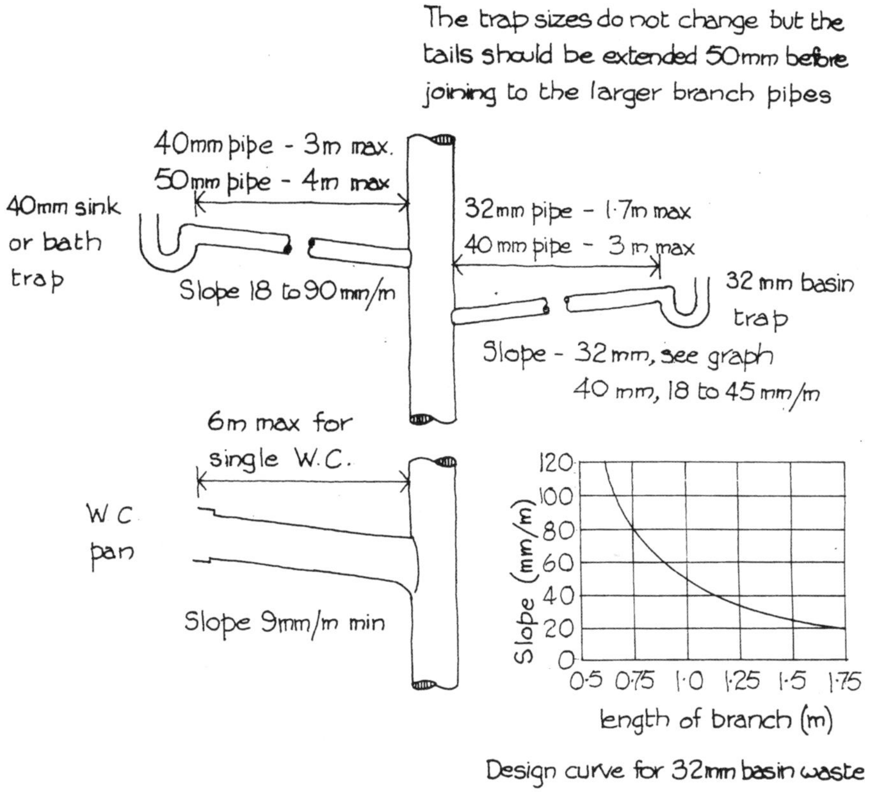

The trap sizes do not change but the tails should be extended 50mm before joining to the larger branch pipes

40mm pipe - 3m max.
50mm pipe - 4m max

40mm sink or bath trap

Slope 18 to 90 mm/m

32mm pipe – 1·7m max
40 mm pipe – 3 m max

32 mm basin trap

Slope - 32mm, see graph
40 mm, 18 to 45 mm/m

6m max for single W.C.

WC pan

Slope 9mm/m min

120
100
80
60
40
20
0

Slope (mm/m)

0·5 0·75 1·0 1·25 1·5 1·75

length of branch (m)

Design curve for 32mm basin waste

Fig. 13.4 Lengths and gradients of branch pipes

The size of the ventilating pipe in a branch serving one appliance should be not less than 25 mm, and where the branch is longer than 15 m or has more than five bends, should be at least 32 mm.

13.3.6   Soil stacks

(Referred to in Approved Document H1 as discharge stacks.) There is no requirement given in the Approved Document as to whether to fit a soil stack inside or outside a building less than three storeys in height. Over three storeys high the soil stack should be located inside the building.

The minimum diameter for a soil stack serving appliances other than a WC is 50 mm if the flow is under 1.2 l/s and 65 mm if the flow rate is between 1.2 and 2.1 l/s. Where there is one siphonic WC only plus any other appliances, and the flow rate does not exceed 3.4 l/s, the stack size can be 75 mm. Stacks of 90 or 100 mm diameter are suitable for all types of appliance, subject to maximum flow rates of 5.3 and 7.2 l/s respectively.

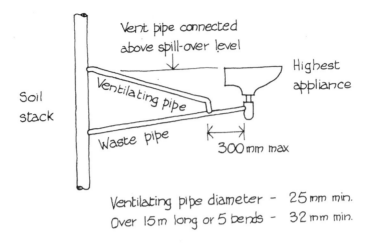

Ventilating pipe diameter – 25 mm min.
Over 15 m long or 5 bends – 32 mm min.

Fig. 13.5 Waste branch ventilating pipe

There should be no offsets in the 'wet' portion of a soil stack, if they can be avoided. Where it is impossible to design the system without an offset, there should be no branch connections within 750 mm of the offset. If the building is more than three storeys in height, a ventilating stack may be needed with connections above and below the offset.

All soil stacks should, of course, discharge to a drain and the radius of the bend at the foot of the stack should be not less than 200 mm to the centreline of the pipe – larger if possible.

13.3.7 Ventilation of soil stacks

Soil stacks should be ventilated for the same reason as long branch pipes, to prevent the build-up of pressures that may force the water out of the seals in the traps. Furthermore, soil stacks connected to drains that are liable to surcharging or backing up, or are connected near an intercepting trap require ventilating pipes of not less than 50 mm diameter connected above the likely flood level of the drain.

That part of the soil stack that serves as a ventilating pipe only – the dry part above the highest connection – should be the same diameter as the wet part of the soil pipe. Soil stack ventilating pipes may terminate outside or inside the building. Those terminating in the outside air should do so at least 900 mm above any openings into the building within 3 m of the pipe and should be finished with a cage or terminal that does not restrict the flow of air. Those terminating inside the building should be fitted with an air admittance valve that has a current British Board of Agrément Certificate. The use of such air admittance valves should be carefully monitored to ensure that they do not adversely affect the amount of ventilation necessary for the below ground drainage system as normally provided by open top ventilating pipes.

13.3.8   Ventilating stacks

This is a dry pipe that provides ventilation to branch pipes as an alternative to carrying them to outside air or to a ventilated soil stack. The lower end of the ventilating stack may be connected to a bend or directly to the soil pipe below the level of the lowest branch pipe. The upper end may be carried to the outside air or it may connect to a ventilated soil pipe above the spill-over level of the highest fitting.

13.3.9   Stub stacks

Unventilated stub stacks may be used for above ground drainage within certain limits. The stub stack should connect either into a ventilated soil stack or directly into a drain. Branch pipes taking waste water only should be connected to the stub stack within 2.0 m of the connection to the ventilated stack or drain. Branches that serve a WC should connect to the stub stack within 1.5 m measured from the invert of the drain or connection to the ventilated stack to the crown of the WC trap.

The length of branch drain from a stub stack should be limited to 6 m where a single appliance is connected and 12 m where a group of appliances are served.

13.3.10   Access

It must be possible to gain access to all parts of the above ground system for the purpose of clearing any blockages that might occur.

Rodding points should be fitted in any length of discharge pipe which cannot be reached by the removal of a trap and also where necessary to give access to any length of pipe which cannot be reached from any other part of the system.

In addition, all pipes should be reasonably accessible to allow any essential repairs to be carried out.

13.4   Macerators and pumps

Sanitary fittings may be connected to a macerator and pump where there is inadequate fall for a gravity system. The discharge from the pump may be taken through a small bore waste pipe to a soil pipe.

The macerator, pump and small bore system must be the subject of a current European Technical Approval issued by a member body of the

European Organization for Technical Approvals, such as the British Board of Agrément and the conditions of use are in agreement with the terms of that Approval Document.

If it is the WC that is connected to the macerator and pump, there must be another WC, connected to a gravity system, accessible in the building.

13.5   Pipes penetrating compartment floors or walls

There is a clear risk of fire penetrating from one fire-resisting compartment of the building to an adjoining compartment via the holes left in the enclosing walls and floor if the soil or waste pipes penetrating them collapse when subjected to heat. Alternatively, fire may escape if the opening provided to allow the passage of a fire-resisting pipe is larger than the pipe, thereby leaving a gap through which flames can penetrate.

The first of these problems is dealt with by restricting the use of combustible pipes and the second is treated by fire stopping round the pipe or a proprietary fire-sealing system, alternatively the pipes may be enclosed in a duct or sleeve pipe. This is covered in detail in Section 6.5 of Chapter 6 with respect to the penetration of compartment floors and Section 8.4.3 of Chapter 8 with respect to compartment walls.

13.6   Alternative approaches

As an alternative to the detailed recommendations given in the Approved Documents and set out above, the requirements of the Regulations can be met by adopting the relevant recommendations of the following British Standards:

BS 6465: *Sanitary installations:* Part 1: *Code of practice for scale of provision, selection and installation of sanitary appliances.*

BS 5572: *Code of practice for sanitary pipework.*

BS 8301: *Code of practice for building drainage.*

Chapter 14

BELOW GROUND DRAINAGE, RAINWATER DISPOSAL AND SOLID WASTE STORAGE

14.1   The Building Regulations applicable

The Building Regulations relating to the disposal of the waste products of a building are:

B1   Means of escape
B3   Internal fire spread (structure)
E1   Airborne sound (walls)
H1   Sanitary pipework and drainage
H2   Cesspools and tanks
H3   Rainwater drainage
H4   Solid waste storage

The relevant part of Regulation B1 is concerned with the positioning of refuse chutes in relation to escape routes from buildings; B3 deals with the potential fire risk created where a refuse chute penetrates a compartment wall or floor; E1 sets satisfactory sound insulation standards for walls separating habitable rooms or kitchens in a building from another part not used as part of the dwelling, in this case the particular requirement relates to the insulation needed in a wall enclosing a refuse chute; H1 requires that any system that carries foul water shall be adequate; H2 states that cesspools and the like shall be of adequate capacity and correctly constructed and sited; H3 requires that any system that carries rainwater shall be adequate and H4 calls for a similar adequacy in the storage of solid waste.

While the Regulations require that systems carrying rainwater shall be adequate, no mention is made of any of the methods of disposal. Reference is made, however, to BS 8301: *Code of practice for building drainage* and the recommendations contained in this Standard are quoted in 14.10.

In addition to the requirements of the Building Regulations on the subject of solid waste disposal, guidance can be found in BS 5906: *Code of practice for storage and on-site treatment of solid waste from buildings; The Special Waste*

Regulations 1996 published by HMSO as part of Statutory Instrument 972 1996 on Environmental Protection and in *A strategic guide to clinical waste management*, published by NHS Estates. Where relevant, information given in these publications is included in this chapter.

14.2   Drain design and lay-out

The design of the drainage system depends on whether it will discharge into a public sewer or an on-site disposal method and, if there is a public sewer, whether it is for soil drainage only or a combined sewer taking both soil drainage and rainwater. The latter is less common at the present time but if it is available, the pipes for the drainage must be sized to take account of the possibility of the peak flow rates of both the soil drainage and the rainwater drainage occurring at the same time. If the discharge is to a cesspool or septic tank a combined system cannot be used.

The normal, and preferable, arrangement is to design the system to fall all the way to the public sewer, but if such a gravity system is not possible or impracticable, sewage lifting equipment would have to be employed and should be installed in accordance with the recommendations contained in BS 8301: *Code of practice for building drainage.*

The connection between drain runs and the sewer should be made obliquely and in the direction of flow.

The simpler the lay-out of the system the better it will work, changes of direction and gradient should be as few as possible and as easy as practicable. Access points should be provided only where they are necessary to allow a blockage of the drain to be cleared.

The system needs to be ventilated by a flow of air and a ventilating pipe should be provided:

- at or near the head of each main drain
- at the head of any branch longer than 6 m which serves a single appliance
- at the head of any branch longer than 12 m serving a group of appliances
- on any drain fitted with an intercepting trap, particularly in a sealed system

Pipes should be laid to even gradients and in straight lines as far as practicable. Where essential, bends of as large a radius as practicable may be used in the drain run, provided that any possible blockages can still be cleared and they should be restricted to a position close to an access point or at the foot of a soil pipe.

14.2.1   Materials for pipes and jointing

Any of the materials that comply with the British Standards set out below are acceptable:

Material		British Standard
Rigid pipes:	asbestos	BS 3656
	vitrified clay	BS 65, BSEN 295
	concrete	BS 5911
	grey iron	BS 437
Flexible pipes:	uPVC	BS 4660 and BS 5481

The joints should be formed in material appropriate to that of the pipes and, to minimize the effect of differential settlement, all joints should be flexible. Nothing should project into the drain whereby a blockage could be created.

14.2.2 Watertightness

When the drain has been laid, haunched or surrounded and backfilled up to 300 mm it should be tested by a water test or an air test.

The finished drain should be capable of withstanding a water test pressure produced by 1.5 m head of water above the invert of the pipe at the head of the drain. If the test is applied by a stand pipe of the same diameter as the drain, it should be filled, left for 2 hours and then topped up. The leakage over the next 30 minutes should then be measured and, in a 100 mm drain, it should not exceed 0.05 litre for each metre run, which is equivalent to a drop in water level of 6.4 mm per metre, or, in a 150 mm drain, it should not exceed 0.08 litre, which is equivalent to a 4.5 mm/m drop. With a long run of drain the amount of the fall added to the test head can induce a pressure capable of causing damage. To prevent this, the test may need to be applied to sections of the drain so that the maximum total head does not exceed 4 m.

An air test is applied using a manometer. This is, principally, a U-tube, graduated in millimetres, containing water. As air is pumped into the drain, the water level in the U-tube rises to indicate the pressure, and pumping continues until the water level reaches either the 50 or the 100 mm mark. This is left for 5 minutes after which time the water level should not have dropped more than 25 mm below the 100 mm head or 12 mm below the 50 mm head.

The inspection chambers should also be tested for water leakage and infiltration. Any tests should be left until the structure of the inspection chamber has had time to attain sufficient strength to withstand the pressures exerted. This is about 3 to 7 days for brick or concrete chambers.

The backfilling should be left out until the test is complete, the drains plugged and the chamber filled with water. After having been left for 20 hours, to allow for any absorption, the level should be topped up and recorded. The test is then to measure any fall in water level over the next 30 minutes; this should not exceed the figures given in Table 14.1.

Infiltration into the chamber should be checked by examination after the excavation has been backfilled. Any visible leaks should be repaired and, generally the rate of any infiltration of ground water should not exceed 0.1 l/m^2/h over the internal surface.

Table 14.1   Maximum drop in water level permitted in manhole tests

Age of manhole	Maximum permissible water level drop (mm) in:		
	A brick or concrete manhole more than 1 m deep	A brick or concrete manhole 1 m or less deep	A clayware, plastic or any 'one-piece' manhole
3 to 7 days	30	20	10
over 30 days	5	5	5

(Source: Based on BS 8301)

14.3   Pipe sizes and gradients

The drains must be capable of carrying the peak flow anticipated. This peak flow depends on the number and type of appliance discharging into the drain. The peak flow rate is not the total discharge from a group of appliances, because they are seldom used simultaneously, and can be determined from Table 14.2 and Fig. 14.1 by adding up the discharge units for the relevant numbers of appliances and converting this total to a peak flow rate by

Table 14.2   Discharge units of appliances

Appliance	Capacity (litres)	Discharge data		Frequency of use, T (s)	Probability of discharge ($P=t/T$)	Discharge units
		Flow rate (litres/s)	Duration, t (s)			
WC (high level)	9	2.3	5	600	0.008	14
				300	0.017	28
Basin (32 mm waste)	6	0.6	10	600	0.017	3
				300	0.033	6
Sink (40 mm waste)	23	0.9	25	600	0.042	14
				300	0.083	27
Shower	–	0.1	–	–	–	Use flow rate
Spray tap	–	0.06	–	–	–	Use flow rate
Urinal, per stall (automatic flush)	4.5	0.15	30	900	0.033	0.3

(Source: Based on BS 8301: Code of practice for building drainage)

Note: The frequency of use time of 600 seconds is suitable for the light usage to be found in a residential building. The frequency of use time of 300 seconds is suitable for the more regular usage experienced in public assembly buildings.

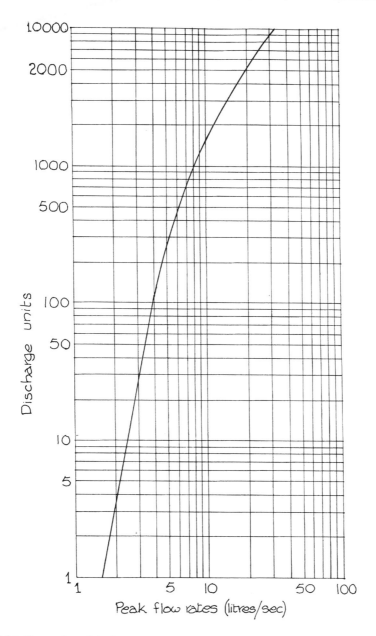

Fig. 14.1 Conversion of discharge units to peak flow rate

reference to the graph in Fig. 14.1. In a combined drainage system it will also be necessary to add the peak flow of rainwater, calculated as described in Section 14.10, before selecting the drain pipe sizes.

The capacity of the drain depends on both the diameter of the pipes used and the gradient at which they are laid. Figure 14.2 shows a graph of the discharge capacities of foul drains running full, and at three-quarters and two-thirds proportional depth. From this graph an appropriate drain size can be selected to accommodate the anticipated peak flow.

While the graph in Fig. 14.2 gives the figures for a 75 mm drain, BS 8301 recommends that, for foul drains, 100 mm is the minimum size of pipe that should be used. The British Standard also states that the flow rate should not

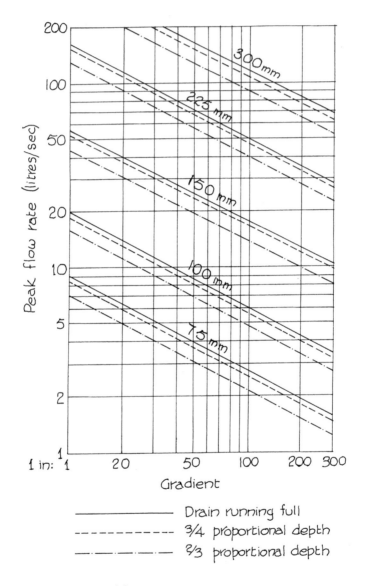

Fig. 14.2 Discharge capacity of drains

exceed the drain capacity when flowing three-quarters full and if the flow is less than 1 l/s the gradient for a 100 mm pipe should not be flatter than 1 in 40. Other gradients recommended is BS 8301 are:

- 1 in 80 for a 100 mm drain where the flow exceeds 1 l/s and there is at least one WC connected to the drain.
- 1 in 130 for a 100 mm drain where a very high standard of design and workmanship can be achieved.
- 1 in 150 for a 150 mm drain where at least 5 WCs are connected.
- 1 in 200 for a 150 mm drain where a very high standard of design and workmanship can be achieved.

It should be noted that, although the larger diameter pipes can be laid at a shallower gradient, increasing the pipe diameter beyond that required for the flow, simply to solve a problem with the falls, is not a satisfactory design.

The minimum WC requirement mentioned above is because a drain pipe needs to have an adequate flow through it to keep it clean, particularly at the flatter gradients where the effluent has a low velocity.

14.4 Bedding and backfilling

The choice of bedding and backfilling materials depends on the depth at which the pipes are laid, their size and strength and the nature of the ground excavated.

The last of these is involved where it is possible to re-use the excavated sub-soil material for backfilling purposes. The following on-site test is described in BS 8301 to determine whether the 'as-dug' material is suitable.

Firstly, a visual examination should be made to see whether there are any particles in excess of 20 mm, any more than a very small proportion would render the material unsuitable. If tested by a sieve, a 2 kg sample should be taken and nothing should be retained on a 38 mm sieve with no more than 5 per cent by mass on a 19 mm sieve.

Secondly, a compaction test should be made on a representative sample of about 11 kg. To do this, a cylinder is required, 150 mm in diameter and 250 mm long (a length of uPVC pipe is quite suitable), placed on end on a smooth, flat surface, gently filled with the material poured in until it is level across the top. The pipe is then lifted clear and placed on another smooth, flat surface. One quarter of the material previously in the pipe is returned to it and tamped with a rod, 40 mm in diameter and weighing 1 kg, until the sub-soil is firm. This is followed by the rest of the sub-soil, a quarter at a time, each firmly tamped. The top of the final quarter should be finished off as level as possible. The distance from the top end of the pipe down to the surface of the tamped material divided by the length of the pipe (250 mm) is called the compaction fraction.

If the compaction is 37.5 mm or less – giving a maximum compaction fraction of 0.15, the material is suitable, if it is between 37.5 mm and 75 mm –

Table 14.3   Limit of cover (mm) over rigid drain pipes

Pipe bore	Bedding detail (Fig. 14.2)	Limit of cover in:					
		Fields and gardens		Light traffic roads		Heavy traffic roads	
		Min.	Max.	Min.	Max.	Min.	Max.
100	A or B	400	420	700	410	700	370
	C	300	740	400	740	400	720
150	A or B	600	270	110	250	–	–
	C	600	500	600	500	600	460

(*Source:* Based on Table 8 in Approved Document H1)

giving a maximum compaction fraction of 0.30, the material may be suitable in normal ground conditions with care but it would be unsuitable if there was a possibility of the trench becoming waterlogged. Material with a compaction fraction over 0.30 is not suitable for backfilling.

The appropriate granular material for backfilling and bedding is determined by the pipe size as follows:

Pipe size (mm)	*Bedding material*
110	10 mm nominal single sized aggregate
160	10 or 14 mm nominal single sized aggregator, or 14 to 5 mm nominal graded aggregate
225 and over	10, 14 or 20 mm nominal single sized aggregator, or 14 to 5 mm or 20 to 5 mm nominal graded aggregate

Suitable arrangements for bedding and backfilling of both rigid and flexible pipes are shown in Fig. 14.3. In detail A, the barrel of the pipe should be supported by the bottom of the trench, which must be very accurately trimmed to the falls by hand and with suitable pockets taken out to receive the collars. This is only suitable where the trench is in ground that would pass a compaction fraction test.

The minimum and maximum depths of cover for rigid pipes are shown in Table 14.3. For flexible pipes the maximum depth of cover should be 10 m and the minimum 0.9 m under any road or 0.6 m under any other surface. In both cases the minimum can be reduced if special protection is provided.

14.5   Special protection

There are situations where the standards set out in the Approved Document cannot reasonably be achieved or where a particular set of circumstances require extra precautions to be taken. To allow for this the Approved Document describes a number of special protection measures that can be taken.

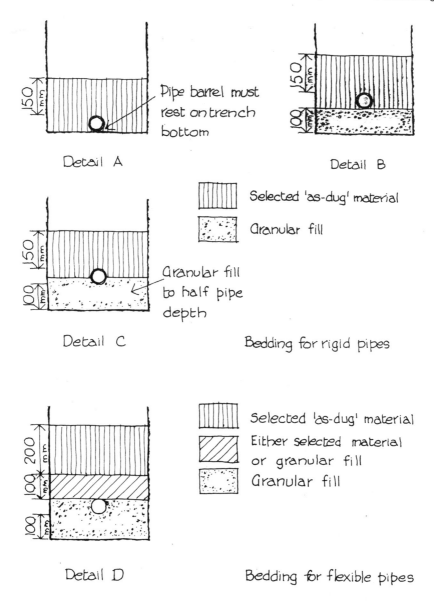

Detail A

Pipe barrel must rest on trench bottom

Detail B

▥ Selected 'as-dug' material

▦ Granular fill

Detail C

Granular fill to half pipe depth

Bedding for rigid pipes

Detail D

▥ Selected 'as-dug' material

▨ Either selected material or granular fill

▦ Granular fill

Bedding for flexible pipes

Fig. 14.3 Bedding for drain pipes

14.5.1 Pipes under buildings

Drains may be run under buildings in the normal way, provided that there is at least 100 mm of granular fill around the pipes unless the crown of the pipe is within 300 mm of the underside of the ground floor slab, in which case concrete encasement, integral with the slab, should be used.

If there is the possibility of excessive settlement occurring, additional flexible joints may be required or, alternatively, the drain could be suspended.

14.5.2 Pipes through walls

To accommodate any possible movement of the wall one of two options should be adopted:

1. An opening is formed in the wall leaving at least 50 mm clearance all round the pipe, with masking sheets fitted round the pipe against both wall faces to close off the opening against possible ingress by vermin.
2. A pipe, built into the wall, is of such a length that both ends are just clear of the face. Flexible jointed rocker pipes, no more than 600 mm long, are connected to each end of the fixed pipe, and these, in turn, are connected to the drain runs (see Fig. 14.4).

14.5.3 Pipes near walls

A drain trench should not be excavated below the level of the foundations of any nearby building unless:

● where it is within 1 m of the foundation, the trench is filled with concrete up to the level of the underside of the foundation structure, or
● where it is more than 1 m from the foundation, the trench is filled with concrete to a point which is lower than the underside of the foundation structure by a height equal to the distance from the edge of the foundation structure to the drain trench, minus 150 mm (see Fig. 14.5).

14.5.4 Reduced amount of cover

Where it is not possible to provide the amount of cover to rigid pipes shown in Section 14.4, the drain should be encased in not less than 100 mm of concrete with movement joints formed with compressible board at each socket.

Flexible pipes with less than 600 mm of cover should be laid as normal, with at least 100 mm of granular fill below the pipe and 75 mm above. This is then protected by concrete slabs laid to bridge the filling (see Fig. 14.6).

14.6 Surcharging of drains

Where a drain is liable to surcharging or flooding, such as during a storm, measures should be taken to protect the building. The precise action to be taken depends on the circumstances, and guidance on protective measures is to be found in BS 8301. Where any type of anti-flood device is installed, it may be necessary to provide additional ventilation of the drain to avoid the trap seal being lost.

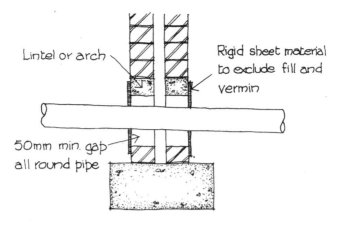

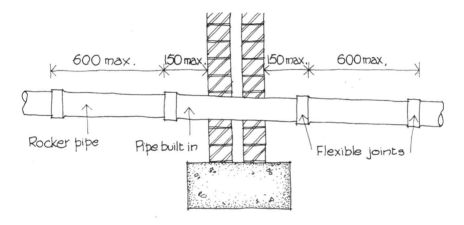

Fig. 14.4 Drain pipes passing through walls

Three protective devices are described in BS 8301:

- A flap valve at the outlet. This relies on a build-up of head on the inlet side to open the flap. Debris lodging on the valve seating or stiffening through lack of use may prevent efficient operation of the flap. Its application is, therefore, limited to surface or clean water drainage.
- A flap and float valve. In this form the float operates in a side chamber, ventilated to atmosphere, and this activates the flap to close the drain.
- An in-line valve. This is recommended as being the best form of protection and can be either manually or automatically operated.

The British Standard also states that the use of any device that relies only on a floating ball is not to be recommended.

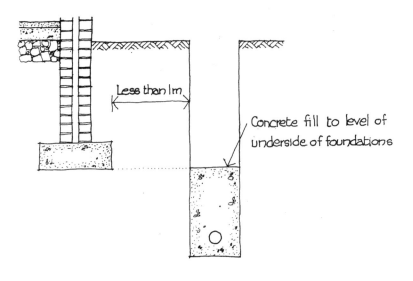

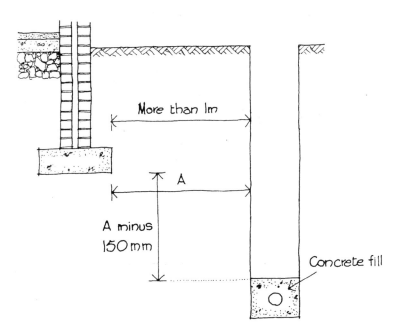

Fig. 14.5 Drains near to building foundations

14.7 Rodent control

In certain areas the infestation of the drains and private sewers can be a problem. Approved Document H1 suggests that the local authority can

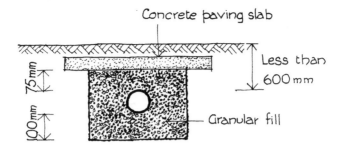

Fig. 14.6 Minimum protection for flexible pipes

provide the best information on the protective measures found to be the most effective in their particular area.

14.8   Access

To make it possible to clear any blockages that might occur in the drains, sufficient and suitable access points are to be provided, of a type sited and spaced appropriate to the lay-out of the system, its depth and the size of the drain runs.

The Approved Document covers the normal method of clearance by rodding and not the use of any mechanical means.

Four types of access points are covered: rodding eyes, access fittings, inspection chambers and manholes. These are defined as:

- Rodding eyes: capped extensions of the pipes.
- Access fittings: small chambers on (or an extension of) the pipes but not with an open channel.
- Inspection chambers: chambers with working space at drain level.
- Manholes: large chambers with working space at drain level.

The depths at which each type should be used and the dimensions recommended are shown in Table 14.4.

14.8.1   Siting and spacing of access points

Access points are required at:

- the head of the drain
- a bend
- a change of gradient
- a change of pipe size
- a junction (unless each run can be cleared from another access point)

273

Table 14.4 Minimum dimensions (mm) for access fittings and chambers

Type	Depth	Minimum internal sizes		Minimum cover sizes	
		Length × width	Circular diameter	Length × width	Circular diameter
Rodding eye		As drain but minimum 100			
Access fitting:					
small	600 or less	150 × 100	150	150 × 100	150
large	600 or less	225 × 100	–	225 × 100	–
Inspection chamber	600 or less	–	190*	–	190*
	1000 or less	450 × 450	450	450 × 450	450†
Manhole	1500 or less	1200 × 750	1050	600 × 600	600
	over 1500	1200 × 750	1200	600 × 600	600
	over 2700	1200 × 840	1200	600 × 600	600
Shaft	over 2700	900 × 840	900	600 × 600	600

(*Source*: Based on Table 9 of Approved Document H1)

*For drains up to 150 mm.
† Clayware or plastic may be reduced to 430 mm to support the cover frame.

In addition, there should be access points in the length of any long runs of drain, as shown in Table 14.5.

14.8.2 Construction of access points

All access points must be capable of containing the contents of the drain both under test and under working conditions. In addition they should be able to resist ground water and rainwater finding their way in.

Materials conforming to the following British Standards are suitable for use as inspection chambers or manholes:

Material		*British Standard*
Clay:	bricks and blocks	BS 3921
	vitrified	BS 65
Concrete:	precast	BS 5911
	in situ	BS 8110
Plastics		BS 7158

Rodding eyes and access fittings should be of the same material as the pipes.

Inspection chambers and manholes with open channels should have the branches fitted so that they discharge at or above the level of the top edge of

Table 14.5   Maximum spacing of access points

From:	Maximum spacing (m) to:				
	Access fitting		Junction	Inspection chamber	Manhole
	Small	Large			
Start of drain	12	12	–	22	45
Rodding eye	22	22	22	45	45
Access fitting:					
150 mm diam	–	–	12	22	22
150 mm × 100 mm	–	–	12	22	22
225 mm × 100 mm	–	–	22	45	45
Inspection chamber	22	45	22	45	45
Manhole	22	45	45	45	90

(*Source*: Based on Table 10 of Approved Document H1)

the channel, and wherever the angle of the branch is more than 45° a three-quarter channel section should be used.

Benching of the main channel and the branches should be at a slope of 1 in 12 and carried up to at least the top of the outgoing pipe. It should be finished at the channel with a rounded edge of at least 25 mm radius.

A non-ventilating cover must be fitted to all inspection chambers and manholes. It must be of a durable material, such as cast iron, cast or pressed steel, precast concrete or uPVC and of a strength suitable for its position.

Covers inside buildings should be mechanically fixed and airtight unless the drain itself has a watertight sealed access cover.

Manholes deeper than 1 m should be fitted with step irons or a fixed ladder.

14.9   Cesspools and tanks

Specialist knowledge is advised in Approved Document H2 in the design and installation of small sewage treatment works and guidance can be found in BS 6297: *Code of practice for design and installation of small sewage treatment works and cesspools.*

A cesspool is an underground storage chamber and its capacity should be sufficient to enable it to hold all the soil drainage of the building until it can be emptied. The Approved Document states that this should be at least 18000 litres or 18 m³ below the level of the inlet.

Septic tanks and settlement tanks provide a form of sewage breakdown by a biological action, thereby permitting a degree of disposal. The capacity of septic tanks and settlement tanks needs to be sufficient to allow time for the breakdown of solid matter in the soil drainage and should be at least 2700 litres or $2.7\,m^3$ below the level of the inlet.

If, as is usual, a cesspool, septic tank or settlement tank is intended to be desludged periodically, using a tanker, it should be sited within 30 m of a suitable vehicle access. It should be arranged so that the emptying, desludging and cleaning can be carried out without any hazard to the occupants of the building and without any of the contents having to be taken through a place of work. Access may, however, be through an open covered space.

14.9.1   Design and construction

Cesspools should be designed so that they are:

- covered with heavy concrete slabs and ventilated
- provided with an access for emptying
- provided with access for inspection at the inlet
- have no openings other than for the inlet, access for emptying, ventilation and inspection

Septic tanks and settlement tanks should be designed so that they are:

- covered with heavy concrete slabs and ventilated or fenced in
- provided with access for emptying, desludging and cleaning
- provided with access for inspection at both the inlet and the outlet

Septic tanks should have two chambers operating in series and the inlet and the outlet should be designed to prevent disturbance of the surface scum or settled sludge. Where the width of the chamber does not exceed 1200 mm the inlet should be via a dip pipe. If the incoming drain is laid to a steep gradient, the velocity of the effluent can cause turbulence; to reduce this, the last 12 m should be laid at a gradient of no more than 1 in 50.

The construction of cesspools and tanks can be in engineering bricks laid in 1:3 cement and sand mortar to a thickness of at least 220 mm; concrete at least 150 mm thick of C/25/P mix as specified in BS 5328; glass-reinforced concrete or factory made in glass-reinforced plastics, polythene or steel provided that they carry a British Board of Agrément Certificate and are installed in strict accordance with both the certificate and the manufacturer's instructions. The Approved Document points out that particular care is needed to ensure the stability of factory-made cesspools and tanks. In this respect, the main hazard is flotation when the tank is empty or low and the surrounding water table is higher than the bottom of the chamber. This can occur at the construction stage or subsequently if the chamber is pumped out for cleaning, and may be particularly acute when the tank itself is of a lightweight construction.

The access to a cesspool or tank should have no dimension less than

600 mm where access is required and the cover should be of a durable material consistent with the corrosive nature of the contents of the tank. All covers should be lockable.

14.10   Rainwater collection and drainage

Approved Document H3 gives guidance on ways to meet the requirements for the drainage of rainwater from roofs over 6 m² and it assumes an intensity of fall of 75 mm/h.

14.10.1   Gutters

The material of a gutter should be of adequate strength and durability, all joints should remain watertight under working conditions and the gutter should be firmly supported without restricting any thermal movement. Any differing metals should be separated by non-metallic material to prevent electrolytic action.

The size of a gutter must be adequate to carry the expected flow at any point in the system. This is calculated in relation to the 'effective area', which relates to the actual area of roof to be drained and whether it is flat or pitched, plus the angle of pitch. For a flat roof the effective area is the same as the actual area. For a pitched roof, the effective area is calculated as shown below:

30° pitch	plan area × 1.15
45° pitch	plan area × 1.40
60° pitch	plan area × 2.00
70° pitch and over	elevational area × 0.5

Using these effective areas the gutter size can be found by reference to Table 14.6. The gutter referred to in this table is half round in section, of a

Table 14.6   Gutter and gutter outlet sizes

Effective roof area (m²)	Gutter diameter (mm)	Outlet diameter (mm)	Flow capacity (litre/s)
6.0 to 18.0	75	50	0.38
19.0 to 37.0	100	63	0.78
38.0 to 53.0	115	63	1.11
54.0 to 65.0	125	75	1.37
66.0 to 103.0	150	89	2.16

(*Source*: Based on Table 2 of Approved Document H3)

length from stop end to outlet less than 50 times the water depth, laid level and with an outlet at one end. A gutter with an outlet at each end can have a length of 100 times the water depth. Where the outlet is not at the end, the gutter size should be selected to suit the larger of the roof areas draining into it.

Outlets with rounded edges have a better flow pattern that will allow the downpipe to be reduced in size.

Gutters laid to fall towards the nearest outlet or which have a capacity greater than that of a half-round section can be reduced in size as recommended in BS 6367: *Code of practice for drainage of roofs and paved areas*, or by following the manufacturer's recommendations, provided that they comply with BS 6367.

Gutters should also be installed so that if there is a rainfall greater than normal, the excess water in the gutter can overflow in such a manner that it will be discharged clear of the building.

14.10.2 Rainwater pipes

As with gutters, the materials used for rainwater pipes should be of adequate strength and durability and firmly fixed without restricting thermal movement (which can be significant in pipes of plastics material). All pipe joints inside a building should be capable of withstanding the airtightness test described in this chapter for underground drainage.

The size of the pipe should be not less than the size of the gutter outlet, and where more than one gutter is connected to a pipe it should have a cross-sectional area at least as great as the combined areas of the outlets.

The discharge of a rainwater pipe may be into a drain, into a gully, into another gutter or onto a drained surface. If the drain into which the pipe discharges is part of a combined system, the connection must be through a trap.

14.10.3 Rainwater drainage

Generally, the materials, design, lay-out and installation of rainwater drains should follow the recommendation for soil drains.

The size of the drain should be sufficient to carry the anticipated flow from the roof plus any runoff from paved areas (although this runoff is not covered by the Building Regulations). The paved area runoff should be calculated on the basis of a rainfall intensity of 50 mm/h.

The minimum size of a rainwater drain is 75 mm and the minimum fall for both 75 and 100 mm drains is 1 in 100. Figure 14.2 shows the discharge capacities of drains running full, three-quarters full and two-thirds full. The selection made should ensure that, at peak flow, the drain does not run more than three-quarters full.

14.10.4 Rainwater disposal

The usual methods of disposal of rainwater described in BS 8301, are to a surface water sewer, to a combined sewer, to a ditch, water course or other body of water or to a soakaway.

Where the disposal is to a surface water sewer there is no requirement to fit an intercepting trap to isolate the rainwater drain, but if it is a combined sewer an intercepting trap is essential as rainwater drainage systems generally are not fitted with trapped inlets.

Disposal to a ditch, stream, river, canal, pond or lake is, of course, only feasible if such a feature is conveniently accessible. Where this method is used, the invert level of the outfall should be approximately 150 mm above the normal water level. If there is the possibility of periodic backflooding and it is not practicable to discharge the rainwater at a higher level, a non-return valve should be fitted (see Section 14.6).

The outfall to a ditch or watercourse should be so formed that it provides protection against local erosion to the bank and damage to itself.

Draining into a soakaway is the most common form of disposal. Before installing a soakaway, the sub-soil and general level of the water table should be investigated to check that the contents of the soakaway can drain into the adjoining ground. BRE Digest 151 describes a simple test for the measurement of percolation into the ground that can provide useful information for the selection of size of soakaway needed. The test uses a 150 mm diameter bore hole, 1 m deep into which water is poured to a depth of 300 mm. The time taken for this to disperse is recorded and the procedure repeated several times to obtain an average value. The borehole is then increased in depth by 1 m stages and the test repeated until the proposed depth of soakaway is reached. The recorded results can then be referred to a chart to obtain a suitable diameter and depth for the soakaway.

An alternative sometimes employed, is to calculate the volume of the soakaway on the basis of the storage of one-third of the rain falling on the areas drained. Approved Document H3 states that the intensity of rainfall for calculation purposes should be taken as 75 mm/h for roof drainage and 50 mm/h runoff from paved areas. BS 8301 recommends that, in ground of low permeability, the storage capacity of the soakaway should be equal to the volume of water resulting from a 12 mm fall on the areas drained.

The soakaway, according to BS 8301, should not be less than 5 m from any building, nor in such a position that the ground below a foundation will be adversely affected.

The simplest form of soakaway is a dug pit filled with hardcore to support the sides and finished with layer of top soil. It is quite a good idea to cover the top of the hardcore with polythene to prevent the top soil percolating down into the hardcore, thereby reducing the storage capacity. A larger, more efficient and more expensive method is to use an unfilled pit lined with honeycomb brickwork or precast perforated concrete rings, roofed over, fitted with an access manhole and surrounded by suitable granular material. In either case, the volume is measured below the invert level of the drain inlet.

14.10.5 Rainwater storage

Some building owners or managers now require rainwater to be collected for use, say, in the maintenance of gardens. BS 8301 state that for such purposes it may be drained to a storage container that should be securely covered, with provision for inspection and cleaning, and provided with an overflow to a nuisance-free outfall. The customary arrangement is to fit rainwater diverters in the rainwater downpipes and connect these to a suitable water butt. When the butt is full, the diverter directs the water down the pipe to whatever the means of rainwater disposal that has been provided, thus satisfying the recommendations of the British Standard.

14.11 Refuse storage

Approved Document H4 states that in non-domestic developments it is essential that the collecting authority is consulted for guidance on:

- The volume and nature of the waste and the storage capacity required, based on the frequency of collection and the size and type of container.
- The method of storage, including any on-site treatment proposed, related to the intended lay-out and building intensity.
- The location of storage and treatment areas and the access to them for men and vehicles.
- Hygienic arrangements in the storage and treatment areas.
- Fire hazards and protection measures.

The Approved Document also refers to the recommendations and data to be found in BS 5906: *Code of practice for storage and on-site treatment of solid waste from buildings.* This states, as an approximate guide, that the output of waste per week from a hotel is 3.0 kg per head of staff and residents. Of this, some 4 per cent, by mass, is composed of empty bottles, 8 per cent is waste paper and card, 33 per cent is food waste and 55 per cent is mixed general waste, similar to household waste.

Because of the wide variety of activities taking place within the range of buildings covered in this book, a general guide cannot be given for any other uses except that of clinical waste from hospitals. Advice on this can be found in *A strategic guide to clinical waste management* published by NHS Estates.

Many of the building uses will, like a hotel, generate a significant quantity of waste paper and card that can be re-claimed if separated and stored. A paper store should be dry and vermin proof and of a size capable of holding up to three weeks' waste. Paper should not be stored for more than three weeks before disposal, to avoid the possibility of vermin infestation (the gestation period for mice is 19 days).

The disposal of the waste from other buildings is generally dealt with by the local authority who will, periodically, collect it from approved containers.

It is preferable to keep containers of any kind in a specially designed chamber that should be well ventilated, minimum height 2 m and large enough to allow at least 150 mm between the containers plus space to manoeuvre them in and out of position. The floor should be not less than 100 mm thick, hard, impervious, smooth, washable and coved up at the perimeter, the walls should be non-combustible and impervious. The chamber should be fitted with a 30 minute fire-resisting door that should be self-closing, unless it opens directly to outside air and is located at vehicle access level, well away from the main entrance of the building.

In some buildings it may be appropriate to provide a refuse chute as a means of conveying rubbish to a refuse container. If the building contains a habitable room, care must be taken to provide sound insulation. Approved Document E states that a wall separating a habitable room or kitchen and a refuse chute must posses a mass of at least $1320 \, kg/m^2$.

As a guide, this mass can be achieved by an in-situ concrete wall 600 mm thick with concrete of a density of $2200 \, kg/m^3$ or a three-and-a-half brick wall, one face plastered, using bricks of a density of $1650 \, kg/m^3$

14.11.1   The Environmental Protection Act 1990

Approved Document H4 makes no reference to this Act, however, both the Act and the Regulations made under it are directly applicable to any non-domestic building use and have been greatly developed recently. Previously, the concern of legislation has been with the control and reduction of pollution due to waste, but now environmental protection and waste minimization are seen as the important factors. For this reason, the emphasis in the latest legislation is on the management of waste rather than on its disposal. Both the Royal Commission on Environmental Pollution and the European Commission have turned their attention to re-use, recycling and better management as a means of waste minimization but it is recognized that such re-use and recycling are rarely appropriate for some waste output, especially clinical waste.

To place the responsibility for waste management and minimization on the managers of buildings, the Act has introduced the concept of duty of care. This states that it is the duty of any person who produces, treats or disposes of waste to take all such measures as are necessary to prevent the escape of the waste and to secure the transfer to an authorized person for authorized transport purposes. A written description of the waste must be provided to enable this authorized person to comply with the duty of care with respect to the escape of waste. An authorized person in this context would be: a waste collecting authority, a holder of a waste management licence or any person registered as a carrier of controlled waste. The term 'controlled waste' is defined in the Act as 'any household, industrial or commercial waste', household waste is taken to include that generated by residential homes, universities, schools or other educational establishments and some hospitals and nursing homes. Commercial waste is that arising from premises used for trade, business, sport, recreation or entertainment and also includes clinical waste.

In addition to the term 'controlled waste', the Special Waste Regulations 1996 define a category of 'special waste'. Much of this is the end result of manufacturing processes but it should be noted that the list of hazardous wastes includes:

- waste from the use of certain fats, grease, soaps, detergents, disinfectants, and cosmetics
- waste from the use of pharmaceuticals
- waste from the photographic industry
- wastes arising from human or animal health care or research whose collection and disposal is subject to special requirements in view of the prevention of infection
- batteries and accumulators

In all cases, where this special waste is to be transported the Regulations lay down a procedure of consignment coding and notes to be observed by both the person disposing of the waste and the person conveying it.

It should also be noted that any site where waste is treated, stored or disposed of requires a Waste Management Licence from the Waste Regulation Authority.

The Environmental Protection Act also points out that both the Health and Safety at Work, etc., Act and the Control of Substances Hazardous to Health Regulations may apply to the disposal of clinical waste.

14.11.2 Volume reduction

There is a considerable risk of fire when large quantities of waste or salvaged material are stored for any length of time. The risk can be reduced by volume reduction and baling of the waste.

BS 5906 describes a number of methods of volume reduction, principally, various forms of compaction, shredding and crushing, baling and incineration. Compaction methods used where the waste is suitable for this treatment use a container in which it is placed and compressed until the container is full. This can take the form of waste rammed into a container or packed into a sack or carton. Larger installations use a compactor that compresses the waste into re-usable containers of from 4 to $12\,m^3$ in volume or a static packer in which the container is approximately $12\,m^3$. The containers are wheeled and require to be handled by a special lifting vehicle for emptying. An exchange system may be operated by the collecting authority. These latter methods are more suitable for the larger premises where the volume of waste makes this an economic option.

Shredding and crushing are methods employed where the nature of the waste lends itself to this form of treatment.

Baling is a good method for dealing with large quantities of paper, particularly for recycling purposes.

14.11.3 Incineration

This is a fairly common method of disposal of waste, particularly clinical waste, but if incineration is considered, regard must be paid to the requirements of the Clean Air Act 1956–1968 and the plant must be authorized under the Environmental Protection Act 1990. Incinerators with a throughput of over 1 tonne/hour require the authorization of Her Majesty's Inspector of Pollution and those with a smaller throughput are authorized by the local authority.

Such authorization will set emission limits for chlorides, particulates, carbon monoxide, organic compounds and heavy metals as well as combustion conditions and monitoring requirements. All emissions must be free from visible smoke, fumes, droplets and offensive odours.

To achieve these standards, specified primary and secondary chamber operating temperatures must be observed, along with gas cleaning equipment and suitable chimney heights.

Volume reductions of the order of between 7 : 1 and 10 : 1 can be achieved by this method as well as sterilizing the waste and making it safe to handle, transport and landfill.

14.12   Alternative approaches

The requirements of the Regulations in connection with soil and rainwater drainage and refuse disposal can also be met by compliance with the relevant recommendations of the following British Standards:

BS 5572: *Code of practice for sanitary pipework.*
BS 5906: *Code of practice for storage and on-site treatment of solid waste from buildings.*
BS 6297: *Code of practice for design and installation of small sewage treatment works and cesspools*
BS 8301: *Code of practice for building drainage.*

Chapter 15

MEANS OF ESCAPE AND FIRE-FIGHTING FACILITIES

15.1 The Building Regulations applicable

The safety of the occupants of a building and the ease with which they can escape from a fire are matters that have received the attention of the law makers from the earliest days of building legislation, following the Great Fire of London. The current coverage of the subject is contained in the Building Regulations under:

B1 Means of escape
B3 Internal fire spread (structure)
B5 Access and facilities for the fire service
K2 Protection from falling

B1 is the main area of legislation on this subject and states that the building shall be designed and constructed so that there are means of escape, in case of fire, from the building to a place of safety outside the building, capable of being safely and effectively used at all material times; B3, as part of its general requirements for fire-resisting structures, sets out the standards for protected shafts containing a staircase used as a means of escape; B5 requires that the building shall be designed and constructed so as to provide facilities to assist fire fighters in the protection of life, and provision made within the site to enable fire appliances to gain access; K2 is involved where a means of escape across a flat roof requires guarding.

The requirements contained in Regulation B1 do not, for obvious reasons, apply to any prison provided under Section 33 of The Prisons Act 1952.

15.2 Other legislation

The relation between the Building Regulations and the Fire Precautions Act is that the fire authority cannot impose further conditions regarding means of escape if the local authority is satisfied that the proposed works comply with the requirements of the Building Regulations, unless there is some aspect that is outside the details required for the purpose of Building Regulation Consent. It should be noted, however, that there are some things that can be required under the Fire Precautions Act that do not appear in the Building Regulations. An example of this is the provision of fire-fighting equipment for use by the occupants.

Not only must adequate provisions be made in the first instance, but they must also be properly maintained and the Approved Documents assume that the building will be properly managed. Failure to carry out such management may result in the prosecution of the building's owner or occupier under legislation such as the Fire Precautions Act or the Health and Safety at Work, etc., Act, coupled with possible prohibition of the use of the premises.

In addition to these Acts and legislation, there are a number of Standards published that are relevant to some of the buildings dealt with in this book. NHS Estates, for instance, publish a series of booklets under the general heading of *Firecode*. These give specific recommendations and principles to be followed in health care buildings of all kinds. The Department for Education and Employment also publish a series under the title of *Building Bulletin*. Number 7 in this series sets out the principles and practice to be followed in educational establishments. Similarly, the Home Office has produced a booklet on the subject of fire safety in buildings intended for sporting use with the title of *Guide to Safety at Sports Grounds*. BS 5588: *Fire precautions in the design, construction and use of buildings*: Part 6: *Code of practice for assembly buildings* describes in some detail measures that should be taken to ensure the safety of persons using such a premises. All these are referred to in Approved Document B and, where relevant, the information given is included in this chapter.

15.3   General principles

Any means of escape should be designed to take into account the form of the building, the activities of the occupants, the likelihood of fire and the potential source and spread of a fire. As this varies considerably with the function of the building, reference is made in the Approved Documents to Purpose Groups. These classify the intended use of the building and are defined in Chapter 1 at Section 1.3. If the building contains a mixture of purpose groups, the means of escape from any part intended for dwelling purposes, or for assembly and recreation purposes, must have their own means of escape independent of that provided for the rest of the building devoted to other purposes. In any other case, where a storey is divided into separate occupancies, each must have its own means of escape that does not pass through any other occupancy. If these come together into a common escape route it should be a protected corridor, or a suitable automatic fire detection and alarm system must be installed throughout the storey.

Since the principal concern is to ensure the safety of the occupants, careful observance of the guidance given in the Approved Documents on the subjects of fire-resisting structures or linings and means of escape is essential. It should be noted that this is the sole concern of the Regulations and Approved Documents. The protection of the property, including the building itself, may require additional measures or higher standards before an insurance company would accept the risk.

It is generally accepted that, normally, a fire only starts in one place in a building and initially it creates a hazard only in the small area where it starts. Subsequently it is likely to spread through the building, often along the circulation routes. The chance of it starting in a circulation area is small provided that the combustible content of such areas is restricted.

There is also less risk that a fire will start in the structure of the building than in items such as furnishings that are not subject to the Building Regulations.

Smoke and noxious gases, not the flames, are the primary danger from a fire. Many plastics in common use such as polyurethane or polypropylene are highly toxic when burning. Not only do the products of this combustion obscure the way to escape routes and exits, they may also actually cause casualties. Measures designed to provide safe means of escape from the building must, therefore, take into account the limitation of the rapid spread of smoke and gases.

In providing fire protection of any kind it should be borne in mind that any measures taken which interfere with the day-to-day use of the premises and cause inconvenience to the occupants are likely to be unreliable due to action being taken to eliminate the inconvenience which, thereby, eliminates the effectiveness of the fire safety provision.

Approved Document B1 lays down no general requirements to install any automatic detection and early warning systems. It does, however, mention that consideration should be given to this subject, particularly in public assembly buildings, institutional buildings and those where residential care is provided. In the latter case there are requirements under other legislation to provide an appropriate means of giving warning of a fire. For hospitals, the guidance given in Health Technical Memorandum 82 *Alarm and detection systems* 1996 should be followed.

Reference is made in the Approved Document to the recommendations contained in BS 5839: Part 1: *Fire detection and alarm systems for buildings.* This Standard points out that fire alarms are required to fulfil several different needs and divides systems into two main groups: Type P and Type L.

Type P detection systems are intended to protect property and are automatic so that warning is given should no person be present and manual so that an earlier warning can be given by anyone who is in the vicinity of the fire. The type is subdivided into two further groups, Type P1 providing protection throughout the building and Type P2 where the protection is in defined parts.

Type L detection systems are for the protection of life and are automatic and manual. Three further subdivisions are made: Type L1 for protection throughout the building, Type L2 in certain defined parts (which should include Type L3 areas) and Type L3, which give protection only to escape routes and are the minimum standard stipulated in Approved Document B1.

15.3.1 Definitions

Escape route: The path travelled by persons from the point where they are when the fire starts to a place outside the building, clear of any danger. This, therefore, covers the distance from the theatre seat, school desk, hospital bed, an elderly person's chair, or anywhere where a person may be when a fire starts to a door leading out of the room; the corridors leading from the door to the outside or an escape stairway; the escape stairway and, finally, the means of getting away from the building once outside.

Occupant capacity: Either the maximum number of people the room or storey is designed to hold, or the number found by dividing the floor area (excluding stairs, lifts and sanitary accommodation) by the floor space factor given in Table 15.1.

The capacity of a sports ground is based on the number of people that can be accommodated in the viewing area, according to the Home Office *Guide to Safety at Sports Grounds*. This is the holding capacity and is found from the packing density of 47 persons per $10\,m^2$ or from the capacity of the entrance or exit systems. This is more fully explained in Chapter 9.

Travel distance: The length of the shortest route which, if it includes a stair, is measured along the pitch line down the centre of the steps. For educational establishments, the travel distance is also related to a speed of travel from a fire of 12 m per minute and an evacuation time of 2.5 minutes to a place of safety.

Width of a door: The dimension across the door – not the clear width between the stops.

Width of an escape route: The dimension at 1.5 m above the floor or stair pitch line, ignoring any handrails.

Width of a stair: The clear distance between the walls or balustrades. Strings protruding less than 30 mm and handrails protruding less than 100 mm may be ignored.

15.4 Means of escape

There is a simple general principle to be followed in the design of buildings in relation to safety and that is that any person confronted by an outbreak of fire should be able to turn away from it and make a safe escape. In hospitals the principle has to change slightly because of the dependency of patients. In these buildings, the aim is to achieve an initial and satisfactory progressive horizontal evacuation away from the fire to a place of safety inside the

Table 15.1 Floor space factors

Type of accommodation	Floor space factor (m per person)
1. Standing spectator areas,* spectator area with bench seating,† bar areas without seating	0.30
2. Exhibition areas including the gangways and circulation	0.40
3. Spectator areas with individual seating*	0.40 to 0.50 or the number of seats
4. School or other educational establishment assembly hall	4.45
5. Assembly hall, bingo hall, dance floor or hall, club, crush hall, venue for pop concerts or similar event, queuing area	0.50
6. School or other educational establishment dining hall and gymnasia	0.90
7. Committee room, common room, conference room, dining room, lounge (other than a lounge bar), meeting room, reading room, restaurant, staff room, waiting room	1.00
8. Restaurant round a dance area	1.10 to 1.50
9. Studio for radio, TV or film recording	1.40
10. Exhibition area excluding the gangways and circulation areas	1.50
11. Skating rink‡	2.00
12. Museum, art gallery, dormitory, workshop	5.00
13. Kitchen, library, office, sales area	7.00
14. Bedroom or study-bedroom	8.00
15. Bowling alley	9.50
16. Bed-sitting room, billiards and snooker room§	10.00
17. Storage and warehousing	30.00
18. Car park	Two persons per parking space

(*Source*: Compiled from Table 1 of Approved Document B1 (1992 edition); BS 5588: Part 6; *Building Bulletin 7*, Department for Education and NHS Estates Firecodes)

* The packing density for a sports ground is, according to BS 5588: Part 6, 47 persons per 10 m^2, which translates as a floor space factor of 4.7 m^2 per person.
† An alternative calculation can be based on 450 mm seat length per person.
‡ The floor space factor for an ice rink is, according to BS 5588: Part 6, 1.2 m^2 per person.
§ The floor space factor for a billiards or snooker room is, according to BS 5588: Part 6, 9.5 m^2 per person.

Notes
1. If there is a mixed occupancy of the building, the most onerous factor should be used.
2. Where the accommodation is other than one of the uses quoted above, a reasonable value based on a similar use may be selected.
3. As an alternative, in any building use where fixed seating is provided, the number of seats can be used to give the occupant capacity.

building on the same level and then a further escape to a suitable place of safety outside the building.

There are two basic principles involved in the design of a means of escape. Firstly, in most situations, there should be an alternative means of escape. Secondly, where direct escape to the open air, clear of the fire, cannot be achieved, it should be possible to reach a place of relative safety, such as a protected stairway, within a reasonable travel distance.

The following are not acceptable as means of escape:

- Lifts, except purpose-designed and installed fire-fighting lifts or evacuation lifts intended for use by disabled people.
- Portable ladders and throw-out ladders.
- Manipulative apparatus and appliances such as fold-down ladders.
- Passenger conveyors and escalators – but measures must be taken to safeguard any person using them at the time of the fire.

There are two elements to an escape route: the horizontal and the vertical. The horizontal escape route, dealt with in Section 15.4, comprises the path crossing the room occupied to the exit door and the corridors leading from the door. The vertical element, dealt with in Section 15.10, is the escape stairways and the protected shafts enclosing them.

While it is desirable to provide two directions in which a person could escape from a fire, it is not always possible or practicable to achieve this. It is quite common, for example, for there to be only one door between a room and a corridor; in this case, the escape distance from the furthest part the room to the door should not exceed the maximum distance shown in Table 15.6 or Table 15.7 for one direction only. In large spaces where this cannot be achieved, two exits from the room must be arranged and the angle between them must be at least 45° (see Fig. 15.3).

A single escape route is acceptable from an area within a storey, or from the whole storey where the travel distance to a storey exit is within the limits shown in Table 15.6 or Table 15.7, provided that no one room in this situation has an occupant capacity of more than 50 people. If the storey is used for residential purposes the occupancy limit of any one room is reduced to 30 persons, and the rule does not apply at all to a storey used for in-patient care in a hospital. The single direction of escape also occurs in many cases at the beginning of an escape route, which is acceptable provided that neither the single direction only part of the route nor the part where there is an alternative direction of escape exceed the limits given in Table 15.6 (see Fig. 15.1).

15.5 Internal escape routes and exits

The first part of any internal escape route is usually unprotected and, as explained above, possibly only one way. It consists of the dist across the room and along any corridors. This should be k

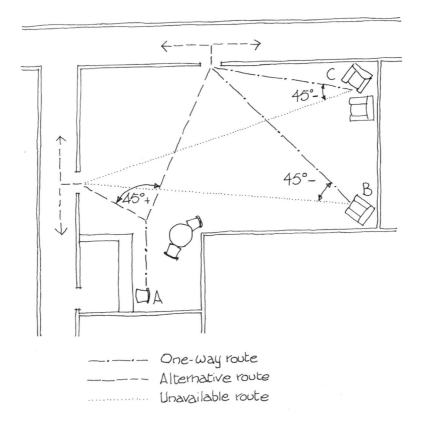

—·—·— One-way route
———— Alternative route
·············· Unavailable route

Fig. 15.1 One- and two-way escape routes

practicable so that people do not have to travel very far while exposed to the immediate danger of fire and smoke, and in any circumstance, should not exceed the distances set out in Section 15.6 below.

At the end of the unprotected escape route there should be either the ultimate place of safety, which is the open air clear of the effects of the fire, a protected corridor or, more commonly, a protected stairway. The length of a protected corridor should be limited because the structure will not provide adequate protection indefinitely.

Protected stairways are designed to provide what is, in effect, a 'fire sterile' area that leads to the ultimate place of safety outside the building. Once they have reached a protected stairway, people should be able to consider themselves safe from the immediate dangers of the fire and can then proceed to a place of safety at their own pace. To achieve this ideal, flames, smoke and gases must be excluded from the protected stairway as far as is reasonable by the provision of a fire-resisting structure or a smoke control system, or both.

Not all stairways must be protected, 'accommodation stairs' can be installed for regular daily use but their role in terms of fire escape is only very limited.

The principle in a hospital, however, is that all stairways will be used for fire escape purposes and must be designed accordingly.

An escape route that crosses a balcony or flat roof should be guarded to prevent anyone falling. The means of guarding is the same as that to be provided for landings to stairs or ramps and should comply with the requirements of Regulation K as set out in Chapter 10 (but see Section 15.9 below).

An isolated step can present a hazard in an escape route as it may cause a fall and, for this reason, it is not permitted except across a doorway.

The floor finish selected for an escape route, including that on any stairs, steps or ramps, must be chosen from material that will not become slippery if it is wet as a result of any fire-fighting operations.

Final exits must be clearly defined and apparent to anybody needing to use them, particularly if the stair continues up or down past the exit. The width of the final exit should be not less than the width of the escape route it serves and located in a position that will ensure the rapid dispersal of persons from the building. Direct access to a street, passageway or open space should be available and the route away from the building well defined and, if necessary, guarded. Final exits should also be located clear of any basement smoke vents, openings to transformer chambers, refuse chambers, boiler rooms or any similar high risk areas.

In a building intended for assembly purposes, BS 5588: Part 6 recommends that all escape routes should:

- Lead directly to a final exit or go by way of a protected space or protected corridor (see Fig. 15.2).
- Have any alternative routes planned to ensure that they do not pass through the same protected space.
- Ensure that if the route contains any ancillary accommodation such as plant rooms, storerooms, kitchens and staff canteens, changing or dressing rooms, stage areas, workshops, loading bays and scenery docks, offices or control rooms then:
  (a) it must be separated by fire-resisting construction of suitable stability
  (b) other measures such as sprinklers, smoke control or fire shutters operated by the smoke detectors must be installed to ensure that the escape route stays open
  (c) exits must be uniformly distributed at intervals of 32 m maximum

There is an alternative to (b) above and that is to separate all escape routes by fire-resisting construction and to size them so that if one is discounted, the rest are sufficiently adequate to enable the total evacuation of the occupants to take place. If this alternative is employed, only the high fire risk areas such as kitchens need to have a fire-resisting construction.

Children in assembly buildings present a special proble~ arrive with a parent or other adult but may, subsequently, pu

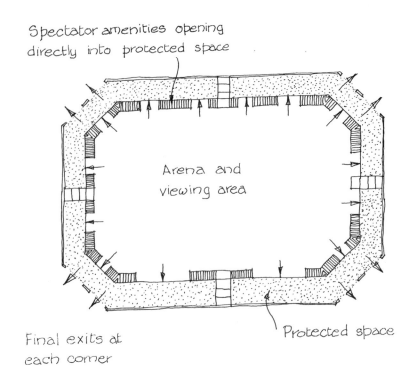

Fig. 15.2 Escape from sports arena via a protected space

another part of the building, separated from their parent. If this is the intention, this separate accommodation for the child's activity should:

- be at or near to the ground level
- never be on a floor above that likely to be occupied by the parent
- never be in a basement storey unless the child's activity room is adjacent to that of the parent
- be adjacent to an external wall
- be supplied with at least 2 fire exits, one of which should be a final exit

Where the parents' and children's areas adjoin, the width of the escape routes and doors should be sufficient for the total number of adults and children.

As has already been mentioned, the principle for escape routes in a hospital is a progressive horizontal evacuation, first to a place of safety within the building on the same level (usually an adjoining compartment) and then a further stage of escape to a place of safety outside the building. To achieve this, the NHS Estates' Health Technical Memorandum 81 states that each compartment within the hospital should be designed to accommodate both its normal occupants plus the designed capacity of the most highly occupied adjoining compartment. Additionally, certain limits on the numbers accom-

Table 15.2 Conditions applying to multi-storey hospitals

Height of storeys above the level that contains the patient access areas (m)	Area of the storey (m²)	Conditions
Up to 12	Less than 1000	No more than 30 patients per storey Each storey divided into at least two compartments If sleeping accommodation is provided there should be no more than 20 beds per storey
Up to 12	More than 1000	As above but each storey divided into at least three compartments
More than 12	Any area	Each storey divided into at least four compartments Each compartment at least 500 m²

(*Source*: Health Technical Memorandum 81: *Fire precautions in new hospitals*, NHS Estates)

modated and other conditions are imposed where the building is of a multi-storey design. These are set out in Table 15.2.

15.5.1   Exit signs

All escape routes and exits must be distinctly and conspicuously indicated by an exit sign. Approved Document B1 states that these must be in accordance with BS 5499: Part 1. This is only an interim standard applicable until the European Safety Signs Directive takes effect. The British Standard replaced the older BS 5260 sign (which just simply displayed the word EXIT) with a sign showing the international pictogram of a man running through a door. This pictogram is to be accompanied by a supplementary sign bearing the word 'EXIT' or the words 'FIRE EXIT' or 'EXIT for emergency use only'. The signs are to be in green and white, using a green background with a green figure in a white doorway for face illuminated or non-illuminated signs and a white man in a green doorway for self-luminous or internally illuminated signs.

The pictogram can show the figure running either to the right or to the left. A sign with the figure to the right and a supplementary sign with an arrow indicating right shows that the escape route changes direction to the right. Without the arrow, the sign indicates that there is no change of direction. The figure running to the left shows that the route has a change of direction to the left.

Both the Exit sign with its pictogram and the adjoining supplementar··· or signs must either have their adjacent edges the same ··· combined onto one signboard.

The European Safety Signs Directive dispenses with an

pictogram of the running man, an arrow and a rectangle, indicating a door. Figure 15.3 shows the old BS 5260 'EXIT' sign, the BS 5499: Part 1 sign and the new sign prescribed by the European Signs Directive. The Directive also requires that internally illuminated signs must be visible from 200 times the height of the panel and separately illuminated signs from 100 times their height. This means that, for the same viewing distance, separately illuminated signs will have to be twice the height of an internally illuminated sign.

While compliance with the Regulations requires the BS 5499 sign to be used, it is intended that the European Directive will have a retrospective

The old BS 5260 sign

The current BS 5499: Pt 1 sign

The new European Directive sign

Fig. 15.3 Exit signs

application to all existing buildings and it would be prudent, therefore, to install this type from now on.

15.5.2 Number, width and height of escape routes

The basis for deciding the minimum number and width of escape routes and final exits is, nearly always, the anticipated occupancy of the room, area or storey. The way to calculate the occupant capacity is given in 15.2.1 above. Using this basis, the minimum number of escape routes and their widths can be found from Table 15.3, according to Approved Document B1 for general purposes and Table 15.4 for educational buildings as laid down in *Building Bulletin 7* by the Department for Education.

For hospitals, the minimum width in a 'hospital street' (see 15.4.3) is 3 m, in departments where beds and patients are being moved the width should be adequate for these activities but elsewhere the minimum width of escape routes, based on the number of people using them should be:

- for up to 200 people: 1100 mm
- for over 200 people: an additional 275 mm for each 50 persons over 200

The number of routes in a hospital is either two or three, as shown in Table 15.2, and limits are placed on the travel distances as given in Section 15.6.

In an assembly building a single escape route is permitted where it is from a

Table 15.3 Number and width of escape routes for buildings generally

Maximum number of persons	Minimum number of escape routes or exits	Minimum width (mm)
50	1*	800†
110	2	900
220	2	1100
500	2	1250
1000	3	1667
2000	4	2500
4000	5	4000
7000	6	5833
11000	7	7857
16000	8	10000
More than 16000	8 plus 1 per each 5000 persons over 16000	5 mm per person in each escape route

(*Source*: Tables 4 and 5 of Approved Document B1)

* Subject to the length of the escape route (see Table 15.6).
† May be reduced to 530 mm for gangways between fixed storage ra areas.

Table 15.4   Number and width of escape routes in educational establishments

Maximum number of persons	Minimum number of escape routes or exits	Minimum width of each escape route or doorway (mm)
Up to 60	1	726
60 to 100	2	
100 to 200	3	
200 to 300	4	
100 to 213	2	1124
213 to 273	2	1424
273 to 327	2	1724
327 to 387	2	2024
387 to 426	3	1124
426 to 546	3	1424
546 to 639	4	1124
639 to 654	3	1724
654 to 774	3	2024
774 to 819	4	1424
819 to 852	5	1124
852 to 981	4	1724
981 to 1065	6	1124
1065 to 1092	5	1424
1092 to 1161	4	2024
1161 to 1308	5	1724
1308 to 1548	5	2024

(*Source*: Based on Table 1 of *Building Bulletin 7*, Department of Education)

Notes
1. With two or three exits, each exit should lead by a separate route to a final exit. With four or more exits, one exit may use part or the whole of the escape route from one of the other exits.
2. All the figures given are based on the assumption that one of the exits or routes is blocked.
3. The values are derived from an evacuation rate of 40 persons per minute for 2.5 minutes and an exit or route width of 530 mm per person.
4. The number and width of escape routes given apply to assembly halls when used solely for educational activities. If they are to be used for public assembly, a licence will be required and the local licensing authority may require a higher standard.

room only (not a storey), provided that the room does not accommodate more than 50 persons and the travel distance does not exceed 15 m, if there are seats fixed in rows, or 18 m if it is an open area. In any other location the

Table 15.5   Minimum widths of escape routes in assembly buildings

Maximum number of persons using the escape route	Minimum width (mm)
50	900
110	1000
220	1100
240	1200
260	1300
280	1400
300	1500
320	1600
340	1700
360	1800

(*Source*: Based on BS 5588: Part 6)

Notes
1. Where more than three exits are provided, the capacity must be not less than the number of occupants in the room, tier or storey.
2. Where the number of exits is three or less, each in turn must be discounted when calculating the capacity of the others.
3. Intermediate widths greater than 1100 mm can be calculated by linear interpolation at the rate of 5 mm per person.

minimum number of escape routes in an assembly building, as stated in BS 5588: Part 6, is:

- where the accommodation is for between 1 and 600 persons: 2 routes
- where the accommodation is for more than 600 persons: 3 routes

and the minimum widths are as shown in Table 15.5.

In educational establishments, a standard of exit width of 530 mm per person, a flow rate of 40 persons per minute and an escape time of 2.5 minutes are combined to calculate the minimum width of escape routes, subject to a minimum of two unit widths. For example; a corridor of 1600 mm width would be required if 300 persons are expected to use it as an escape route:

$$\frac{300}{40 \times 2.5} \times 530 = 1590 \, \text{mm}$$

More than one exit must be provided from a room in an educati building when:

- the room is an assembly room, a dining room o accommodate more than 60 persons, or

- the dimensions of the room are such that a part of the room is more than 12 m from the exit, or
- if the room opens into a room or area which is a high fire risk; in this case the second exit must lead, by a separate route, to a final exit, or
- if the room is a laboratory or other high risk area and a single exit would be in a hazardous position, or
- if the room is a lecture theatre seating more than 60 on fixed seating.

The final number of escape routes in any building may well exceed the minimum either because of practical considerations or because of the need to limit the travel distance.

The width of escape routes and exits may also need to be greater than the minimum to allow for the possibility that not all of them may be accessible in a fire. Approved Document B1 states that to allow for this, in any storey where there are two or more exits, the largest must be discounted when calculating the total width required in accordance with Table 15.3. This may also affect the width of any stairways since they must be at least as wide as any exits leading to them.

If an escape route from a ground or basement storey also forms the exit route from an escape stairs the width must be increased accordingly to accommodate the greater number of persons.

Where it is possible that a disabled person may need to use the escape route, the minimum width must then be determined in accordance with Regulation M2. If the corridor is accessible to someone in a wheelchair, the minimum is 1200 mm wide to allow space for manoeuvring and to permit other able-bodied persons to pass. If it is approachable only by a stairway, a lesser minimum width of 1000 mm would be permitted (see Chapter 16).

To ensure that, if one escape route is blocked, the alternative will be available, Approved Document B1 requires that they shall be either at an angle of at least 45° to each other (see Fig. 15.1) or, if less than 45° apart, separated by a fire-resisting construction.

The height of an escape route should be not less than 2 m, except in a doorway.

If the building is designed with a central core containing stairs, lifts and perhaps other facilities such as sanitary accommodation, it is possible that, where required, two or more independent escape routes can exist within it. For this, separately enclosed stairways are required and the exits from the accommodation must be so planned that they are remote from each other and are neither approached from, nor linked by, any lift hall, common lobby or undivided corridor (see Fig. 15.4).

15.6 Travel distances

The maximum distances a person should have to travel to a storey exit are shown in Table 15.6 for assembly and residential buildings according to

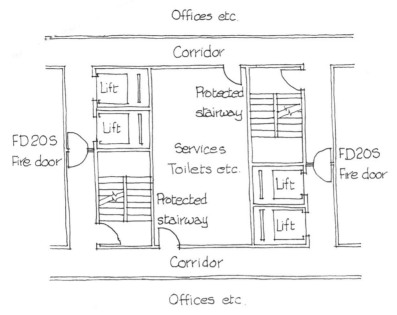

Fig. 15.4 Exits in a central core

Approved Document B1. Table 15.7 shows travel distances for assembly buildings, hospitals and educational buildings according to published standards relating to those building uses, as listed below the table. Two figures are quoted: one for the distance in one direction only and one for the distance where there are two directions of escape.

As already mentioned, the first part of the escape route is across the space occupied and is probably one way only. Referring to Fig. 15.1, it can be seen that escape from easy chair 'A' follows a one-way route for the first leg and then splits into two alternative directions which are more than 45° apart. In this case, the length of the first section must not exceed the 'one direction only' figure given in Tables 15.6 or 15.7 and neither of the second legs must exceed the distance (including any travel distance outside the room) given for the 'more than one direction' route. Escape from both easy chair 'B' and easy chair 'C' are one way only as far as the exit door from the room because the angle formed between the two possible routes is less than 45°.

When the lay-out of any internal partitions, fixed furniture, fittings, etc., is known in advance, the travel distance can be worked out accurately, but where, as is not uncommon in the early stages of the design development, this information is not available, the direct distance between the point under consideration and the exit door can be used. This direct distance must be more than two-thirds of any of the travel distances given in Tab Table 15.7.

Table 15.6 Limitations on travel distance generally

Building use		Maximum travel distance (m) in which travel is possible in:	
		One direction only	More than one direction
Institutional	For hospitals, see Table 15.7	9	18
Residential:			
Bedrooms	Maximum part of travel distance within the room	9	18
Bedroom corridors		9	35
Elsewhere		18	35
Assembly and recreation:			
Buildings primarily for the handicapped, except schools	For schools, see Table 15.7	9	18
Elsewhere		15	32
Storage and other non-residential		18	45
Plant rooms or roof-top plant	Distance within the room	9	35
	Escape route not in the open air*	18	45
	Escape route in the open air*	60	100
Places of special fire risk:			
Oil-filled transformer rooms Switchgear rooms Boiler rooms Fuel stores Stores for highly inflammable materials Any room housing a fixed internal combustion engine	Maximum part of the travel distance within the room	9	18

(*Source:* Based on Table 3 and Appendix E of Approved Document B1)

* Overall travel distance.

Note: The maximum travel distance given is the actual length of the route taking into account any internal partitions, fittings, equipment, etc. If the internal layout is not known when the plans are deposited for approval, a direct distance equal to two-thirds of the dimensions given may be used.

Table 15.7 Limitations on travel distance in assembly buildings, hospitals and educational buildings

Building use	Location	Escape in one direction only (m)	Escape in more than one direction (m)	
			Total distance	Which may include one-way travel up to:
Assembly buildings	Within a storey or tier of seating from:			
	Public areas with seating in rows	15	32	15
	Public open floor areas	18	45	18
	Within ancilliary accommodation:			
	Plant rooms Changing or dressing rooms Fuel stores	9	18	9
	General store rooms Kitchens and staff canteens Stage areas Workshops Loading bays and scene docks Offices Control rooms Any other ancilliary accom- modation	18	45	18
Hospitals	From one compartment to another	64	–	–
	From one sub-compartment to another	15	32	–
	Within plant spaces:			
	Rooms containing ventilation equipment, calorifiers or water storage	15	35	–
	Rooms containing boilers, standby generators, lift motors, batteries, refrigeration equipment or medical gases	12	25	–
	Electrical switchgear rooms	6	12	–
Educational buildings	From any point within a storey:			
	Via a protected two-way route		42	12*
	Via an unprotected two-way route		30	12*
	Via any one-way route	30	–	18*
	From any point within a gallery access storey	–	24	12*
	Across a gallery	15	–	–

(*Source:* Based on BS 5588: Part 6; Health Technical Memorandum 81; NHS Es⁺ *Buildings Bulletin 7*, Department for Education)

* Maximum travel distance within the room.

15.7 Corridors

Most escape routes pass along a corridor to reach the exit to a place of safety. In some circumstances this may need to be a protected corridor, i.e. one that is enclosed by a fire-resisting construction. Alternatively, the enclosure may simply divide the space and provide some defence against the spread of smoke in the early stages.

15.7.1 Means of escape which is not a protected corridor

As mentioned above, the enclosure to a corridor used as an escape route can be considered to afford some defence against the spread of smoke even if it does not possess recognized fire-resisting standards. To maintain this defence, however, the enclosing partitions should be carried up to either a suspended ceiling or the soffit of the structural floor above. Since the structure does not have any inherent fire resistance, any doors between the corridor and the adjoining space do not need to be fire resisting either.

15.7.2 Protected corridors

The corridors to which this applies are:

- every dead-end corridor
- any corridor common to two or more different occupancies

The construction of an internal fire-resisting wall is described in Chapter 8 at 8.4. Generally, the standard of fire resistance to be achieved is 30 minutes in respect of its loadbearing capacity, its integrity (its resistance to the fire breaking through) and its insulation (the length of time it can resist the transference of excessive heat). A limited amount of glazing is permitted in these walls as described in Chapter 8 at 8.7.5.

Where the protected corridor is on the top floor, the part of the roof that encloses it must possess the same fire-resisting rating.

15.7.3 Subdivision of corridors

Every corridor, more than 12 m long, that connects two or more storey exits must be subdivided by self-closing fire-resisting doors so that:

- no length of undivided corridor is common to two storey exits
- the fire doors are positioned to protect the escape route from smoke, having regard to the lay-out of the corridor and to any adjacent fire risks

This is to avoid the possible situation where a corridor, connecting two exits, fills with smoke, making both escape routes impassable before the occupants have had a chance to get out.

Where a dead-end corridor leads to a point where there are two alternative routes for escape, there is a risk that smoke could make both of these routes impassable before the occupants, who have to use the dead-end corridor, have had time to escape. This can be avoided by the installation of a pressurizing system that prevents the build-up of smoke but, failing this, every dead-end corridor over 4.5 m long should be separated by a self-closing FD20S fire door, from any part of the corridor that:

- provides two directions of escape
- continues past one storey to another

This is illustrated in Fig. 15.5 and the grading of FD20S is explained in Chapter 9.

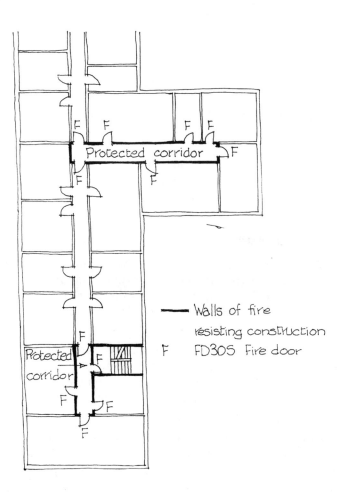

Fig. 15.5 Dead-end corridors

15.7.4 Doors on escape routes

Doors on escape routes must be readily openable as any undue delay in negotiating them may be critical in escaping. In general, these doors should either not be fitted with any form of fastening or fitted with a simple form of fastening, released from the side approached by people making an escape, without requiring the use of a key or the manipulation of more than one mechanism. Panic bolts may be used on final exit doors where security is an important consideration.

Some of the doors in escape routes may need to be fire doors; details of these, plus guidance on self-closing devices, is given in Chapter 9.

All doors on escape routes must be hung to open in the direction of escape to an angle of not less than 90° and clear of any change of floor level other than a threshold or single step on the line of the doorway. If it opens towards a corridor or stairway, the door must be sufficiently recessed to avoid the swing reducing the width of the corridor or landing.

Vision panels are necessary where doors are hung to swing both ways and where they subdivide corridors. Where these corridors are accessible to disabled people in wheelchairs, the vision panel must give a zone of visibility from 900 to 1500 mm from the floor (see Chapter 16).

Revolving and automatic doors should not be placed in escape routes unless:

- they fail safely in the open position, or
- they can easily be opened in an emergency, or
- a swing door or doors of the required width are provided immediately adjacent.

15.7.5 The hospital street

This is a concept widely used in the design of hospitals. It provides a link between the various departments and the stairways and lifts. It also affords a main circulation route for staff, patients and visitors (see Fig. 15.6).

The hospital street is considered as a special type of compartment with the functions of:

- connecting final exits, stairway enclosures and department entrances
- providing the fire-fighting personnel with a bridgehead
- providing a protected escape route from a department for patients if a fire cannot be brought under control within the department

The construction of a hospital street should be to the same standards of fire resistance as the other compartments with a minimum width of 3 m and divided into at least three sub-compartments of a maximum length of 30 m. It should not contain any other accommodation except for sanitary facilities.

If it is on the ground floor it should have a minimum of two final exits, located:

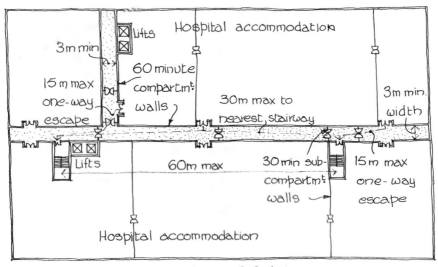

FIRST FLOOR PLAN

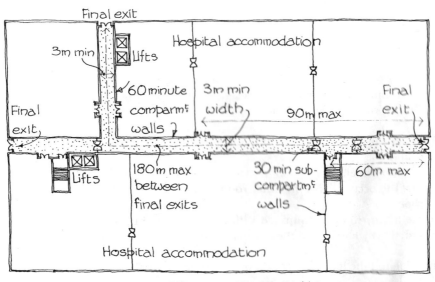

GROUND FLOOR PLAN

Fig. 15.6 The hospital street

- at every extremity of the hospital street
- so that the maximum travel distance between final exits is no more than 180 m
- so that the maximum distance from a compartment exit to a final exit is no more than 90 m

- where at least two of the stairways terminate within 15 m of a final exit which itself is within 18 m of a suitable fire brigade access point

If the hospital street is on an upper floor level the following conditions apply:

- There should be at least two stairways, each in a protected shaft and located so that the maximum distance between them does not exceed 60 m.
- The maximum one-way travel within the street does not exceed 15 m.
- The maximum distance from a compartment exit to a final exit does not exceed 30 m.
- If there are only three sub-compartments in the street, there are stairways in at least two of them and the third is large enough to accommodate all the occupants of the adjoining compartment.

Doorways between the street and adjoining compartments should be arranged so that they are not in the same sub-compartment as the entrances to the lifts or stairs and so that there is always an alternative means of escape from the compartment.

15.8 Inner rooms

An inner room is one that is entered only from another room, known as the access room. This arrangement presents a very real danger if a fire develops in the access room and is only acceptable if:

- the enclosing partitions of the inner room stop at least 500 mm below the ceiling, or
- a vision panel, up to 0.1 m² in area, is provided in the door or partition so that the occupants of the inner room can see if a fire develops in the access room, or
- the access room is equipped with an automatic fire detection and alarm system which will warn the occupants of the inner room of the outbreak.

In addition, the following conditions must be satisfied:

- The access room should be in the control of the same occupier as the inner room.
- The occupant capacity of the inner room does not exceed 50.
- The escape route from the inner room should not pass through more than one access room.
- The travel distance from any point in the inner room to an exit from the access room should not exceed the appropriate distance given in Table

- The access room should not be a 'place of special fire risk' (see Table 15.6).

15.9   External escape routes

Three possible external escape routes are mentioned in Approved Document B1: a simple paved area, an escape route across a roof or external escape stairs. The Document states, however, that a route across a flat roof should not serve an institutional building nor a part of a building to which members of the public will have access. This limitation means that, as far as the buildings within the scope of this book are concerned, the only ones that can make use of such a route are hotels, boarding houses, residential colleges, halls of residence, hostels and like premises and that part of an assembly building accessible by staff only.

Where an external paved area for escape purposes is within 3 m of an external wall of the building, the wall must be of fire-resisting construction up to a height of 1.1 m above the paving level. No specific period of fire resistance is given in the Approved Document.

An escape route across a flat roof is only permissible if there is an alternative means of escape and:

- the route does not serve a building that is intended to be used for residential or educational purposes or a part of the building where members of the public will assemble
- the roof is part of the building from which the escape is being made
- the route across the roof leads to a storey exit
- the part of the roof forming the escape route, its supporting structure and any openings within 3 m of the route are of 30 minute fire-resisting construction in respect of load-bearing, integrity and insulation
- the route is adequately defined and guarded by walls or protective barriers capable of resisting a horizontal force of 0.74 kN/m at a height of 1100 mm above the paving level

External escape stairs, like a route across a roof, are only permissible where an alternative escape route exists and not at all in a hospital. Any such stair must be constructed of materials of limited combustibility and protected from a fire in the building and from the effects of snow and ice if it is more than 6 m to the top floor served. Materials of limited combustibility are defined in 15.10.4 below. The protection from snow and ice does not necessarily require a full enclosure, much will depend on the location of the stairs and the degree of protection afforded by the building.

All doors giving access to the stairs should be fire resisting and self-closing, except where there is only the one exit from the building onto the top landing of the stairs.

The external walls from ground level to 1.8 m above the line of the flights and intermediate landings and 1.1 m above the top landing are to be of fi-

resisting construction. The Approved Document does not specify the period of fire resistance required for this area of wall.

15.10 Escape stairs and protected shafts

As already mentioned, any escape route must terminate in either the open air or a 'fire sterile' protected shaft. Progress from this protected shaft must be via an escape stairs, ultimately leading to the open air. It is important that the number and size of these escape stairs are sufficiently adequate to deal with the numbers of people likely to use them.

The underlying principle in the provision of alternative escape routes is that every person in the building, unless it is very small, is offered another way of getting out should one be blocked. To maintain this principle it is necessary to discount each escape stairs in turn and examine the capacity of the remaining provisions to check whether they are adequate for the numbers of people needing to escape. Two exceptions apply to this need to discount one of the escape stairs; if each one is approached on each storey (except the topmost storey) through a protected lobby, then the likelihood of the stairs not being available is significantly reduced, or if the risk is removed by the stairways being protected by a smoke control system designed in accordance with BS 5588: Part 4.

15.10.1 Basement stairs

A stairway leading to a basement is, because of its position, a higher fire risk than those in other storeys and more likely to fill with smoke and heat. For this reason, the basement should be served by its own independent stairway, if there is only one escape stairs in the building. In buildings with more than one escape stairs, only one of them needs to terminate at ground level, the others can extend down to and serve as escape stairs to the basement storeys provided that there is a ventilated protected lobby or corridor between the stairs and the accommodation at each basement level. BS 5588: Part 6 states that in a building used for assembly purposes, not more than 50 per cent of the upper storey stairs are to be carried down as basement stairs and these are to be separated at each basement level by a protected lobby.

15.10.2 Number of escape stairs

Notwithstanding any of the following rules regarding the numbers of escape stairs, a building with a mixed occupancy, each fulfilling a different purpose, will need independent escape stairs from each area, designed in accordance with the requirements for that purpose group.

All buildings of more than one storey need escape stairs but some may only require one to be safe provided that:

Table 15.8   Standard of stairway provision in a hospital

Number of patient beds on any one upper storey	Number of stairways
1 to 100	2
101 to 200	3
201 to 300	4
301 to 400	5

(Source: NHS Estates' Health Technical Memorandum 81: Fire precautions in new hospitals)

- The building has no floor level more than 11 m above ground level and each storey is such that a single escape route from that storey is permissible.
- The stairway is from a basement only, which is sized and occupied in a way that permits the provision of just one escape route.
- The building is not used as a hospital or for educational purposes where, in both cases, a minimum of two stairways is always applied.

In a hospital the number of stairways required is determined by the number of patient beds on any one upper storey and the requirements, as set out in Health Technical Memorandum 81, are given in Table 15.8

The number of stairways in an educational building is controlled by both the number of people expected to be using them and the travel distance between the stairways and the room doors which should not exceed 30 m if the route is through a protected corridor, or 18 m if it is through an unprotected corridor.

In any other multi-storey building the number of escape stairs will be determined by the number of escape routes deemed necessary.

15.10.3   Width of escape stairs

The minimum width of an escape stairs, according to Approved Document B1, should be:

- not less than the width of the exit giving access to it
- 800 mm if the maximum number of people likely to use the stairs in an emergency does not exceed 50
- the width obtained from Table 15.9 or calculated for total evacuation

The widths given in Table 15.9, derived from Approved Document B1, are exactly the same as those given in BS 5588: Part 6 for Places of Assembly.

If the building is more than 30 m high, the width of the stairs must be either less than 1400 mm or more than 1800 mm with a central handrail. A stairs that is over 1400 mm wide is not used in the middle because escaping people like to stay within reach of a handrail and, therefore, would be inadequate for the

Table 15.9 The capacity of a stair for the purposes of escape from a basement or for the total evacuation of a building

Number of floors served	Maximum number of persons served by a stair of width (mm)								
	1000	*1100*	*1200*	*1300*	*1400*	*1500**	*1600**	*1700**	*1800*
1	150	220	240	260	280	300	320	340	360
2	190	260	285	310	335	360	385	410	435
3	230	300	330	360	390	420	450	480	510
4	270	340	375	410	445	480	515	550	585
5	310	380	420	460	500	540	580	620	660
6	350	420	465	510	555	600	645	690	735
7	390	460	510	560	610	660	710	760	810
8	430	500	555	610	665	720	775	830	885
9	470	540	600	660	720	780	840	900	960
10	510	580	645	710	775	840	905	970	1035

(*Source*: Table 7 of Approved Document B1 (1992 edition))

* If the vertical extent of the stair is greater than 30 m, the width should be either less than 1400 mm or more than 1800 mm with a central handrail.

Note: The capacity of the stair serving more than 10 storeys can be obtained from the formula given in the text.

numbers using it, even if the calculations indicated otherwise. A stairway over 1800 mm wide with a central handrail can be considered as two adjoining flights and the spaces on each side of the handrail considered separately for the purpose of assessing the stairs capacity.

In deciding the minimum width of an escape stairs in buildings where the occupancy of the storey exceeds 50, regard must be paid to whether the escape strategy is total evacuation or phased evacuation.

Total evacuation requires the stairs to be designed to allow all the storeys to be evacuated at once (discounting one stairway where required) and phased evacuation assumes that the first people to be evacuated in a fire will be those of reduced mobility, those people in the storey where the fire originated and those in the storey above. Subsequently, if there is a need to evacuate more people, it is done two floors at a time. Approved Document B1 states that the principle of total evacuation should always be used for buildings used for residential purposes or for assembly or recreation. Since this covers all the building uses included in this book, the principle of phased evacuation is not relevant.

In a hospital, the minimum width of a stairway is 1100 mm. If the number of people expected to use the stairway exceeds 200, the width is to be increased by 275 mm for each additional 50 persons.

Table 15.10   School stairway widths and the maximum number of occupants

Number of storeys above the final exit level	Maximum number of occupants on all upper floors when there are two stairways each of width: (mm)				
	1050	1200	1350	1500	1650
2	260	290	330	360	390
3	300	340	380	430	480
4	340	390	440	500	560
5	390	450	510	580	650
6	430	500	570	650	730
7	470	550	630	720	810
8	510	600	690	790	890
9	550	650	750	860	970
10	600	710	820	940	1060

(*Source*: Based on Table 2 of Department for Education *Building Bulletin 7*)

Notes
1. In buildings with three stairways, the values given in the table should be multiplied by 1.8.
2. The number of occupants can be calculated on the following basis:

Dining rooms and gymnasia	0.90 m² of floor area per person
Assembly halls	0.45 m² of floor area per person
Teaching rooms and any other non-teaching accommodation	The maximum number of occupants intended.

In an educational building, the minimum width is 1050 mm and the designed width should be in accordance with Table 15.10. The figures given in Table 15.10 relate only to the requirements for escape in one direction. It may be necessary to increase these widths for normal day-to-day use when the traffic is likely to be in two directions at the same time.

15.10.4   Calculation of minimum stair width for total evacuation

Minimum stair width is based, of course, on the number of people that can be accommodated. If the precise occupancy level of each storey is not known, it should be calculated as shown in 15.2.1.

The formula for the calculation of total stairs width is given in Approved Document B1 as:

$$P = 200w + 50(w - 0.3)(n - 1)$$

where P is the number of people that can be accommodated, w is the total width of stairs in metres and n is the number of storeys.

It should be noted that the total stairs width derived from this formula is not the total of all stairways added together but the total of the stairways left after discounting each one in turn.

A worked example of this formula is given below:

Number of persons in the building:	2400
Number of storeys:	15
Total rise of the stairs:	4.5 m

$$2400 = 200w + 50 \ (w - 0.3) \ (15 - 1)$$
$$2400 = 200w + 700w - 210$$
$$2610 = 900w$$
$$w = 2.90 \, \text{m}$$

Taking into account the need to discount the availability of one stairway in a fire, the conclusions that can be drawn from this result are:

- Two stairways would each have to be equal to the total, i.e. 2900 mm wide which, being over 1800 mm wide, would require a central handrail leaving a stair width of 1450 mm each side. This exceeds the maximum permissible width of 1400 mm imposed on stairs over 30 m high.
- Three stairways, each of a width of 1450 mm, could be considered but as this is more than the maximum of 1400 mm for a single width it would have to be increased to 1800 mm with a handrail down the middle.
- Four stairways each of 967 mm minimum (say 1 m) would meet all the requirements.

The disposition of either the three or the four stairs selected would be determined by the lay-out of the escape routes within each storey and any particular requirements limiting travel distances.

15.10.5   Construction of escape stairs and ramps

Every escape stairs serving a storey more than 20 m above ground or access level or one within a basement storey must be constructed of materials that are of limited combustibility. In practical terms this means, for stairs, concrete, but the definition of limited combustibility also includes any material with a density of 300 kg/m^3 or more which, when tested to BS 476: Part 11, does not flame and the rise on the furnace thermocouple is not more than 20 °C or any product classified as non-combustible under BS 476: Part 4. Although the construction is to be of limited combustibility, combustible finishes may be applied to the treads.

Helical stairs, spiral stairs and fixed ladders may, according to Approved Document B1, provide the means of vertical escape. Due consideration should be given to the function of the building and the practicality of escape by, say, an elderly or infirm person via a helical or spiral stairs. If they are used, they are subject to a number of conditions:

- Helical and spiral stairs must be designed in accordance with BS 5359:

Stairs, ladders and walkways. Part 2 and the guidance given in Approved Document K as set out in Chapter 10. If they are intended to be used by members of the public they must conform to the specification given in BS 5359 for Type E stairs, which require the geometry generally to follow the details in Chapter 10 but subject to the following dimensions:

rise:	150 mm to 190 mm
going: inner edge	150 mm minimum
centre	250 mm minimum
outer edge	450 mm maximum
twice rise plus going:	480 mm to 800 mm
clear width:	1000 mm minimum

- Fixed ladders, which should be of non-combustible material, are only suitable as a means of escape to plant rooms and the like which are not normally occupied and where conventional stairways are not a practical proposition. They are not to be provided for use by members of the public.

Any ramp that forms part of an escape route should have as easy a gradient as possible and in no case should it be steeper than 1 in 12. Any sloping floor or tier should not slope more than 35° and the construction and dimensioning generally should follow the guidance given in Approved Documents K and M as detailed in Chapter 10.

15.10.6 Protection of escape stairs

Escape stairs must be contained within a protected shaft or fire-resisting enclosure unless the travel distance is very short and the number of people involved very small. The degrees of fire resistance and the construction of the walls enclosing stairs are detailed in Chapter 8.

Access lobbies giving added protection to the stairs are required at every level except the top storey, where:

- the stairway is the only one in the building
- the stairway serves a storey at a height greater than 20 m from the ground
- the building is designed for phased evacuation, which does not apply to any of the building uses covered in this book (see 15.10.2)
- the building is fitted with sprinklers so that the calculation of the total stairway width does not require the application of the discounting principle

A protected lobby should also be placed between a protected shaft and a room containing an oil-filled transformer or switch gear, a boiler room, a fuel store, a store for highly inflammable materials or a room housing a fixed internal combustion engine.

Two or more adjoining escape stairs must be treated as completely independent routes to safety and separated by an imperforate fire-resisting wall.

The exit from the protected stairs enclosure should be as wide as the stairs and lead either directly to the open air or connect to a protected corridor leading to the final exit.

As the protected stairs enclosure must be free from any potential fire risk, the additional facilities that may be incorporated within it are strictly limited to:

- sanitary accommodation and washrooms, provided that these are not used as cloakrooms
- a lift well, provided it is not a fire-fighting stairway
- a reception desk or enquiry office of not more than $10 \, m^2$ at ground floor or access level, provided that there is another escape stairs in the building
- a cupboard, provided that it is enclosed with fire-resisting construction and there is another escape stairs in the building

Gas water heaters and incinerators may be installed in the sanitary accommodation within a protected stairway but gas service pipes and associated meters should not be incorporated within the area unless the gas installation is in accordance with the requirements for installation and connection set out in the Gas Safety Regulations 1972 and the Gas Safety (Installation and Use) Regulations 1984 as amended in 1990.

15.11   Passenger lifts and evacuation lifts

Generally, passenger lifts must be discounted from use in the event of a fire as a failure of the electrical system could immobilize the lift and trap anyone within it, but the lift shaft and protection of the lift itself must be carefully considered since it connects floors and probably penetrates the compartmentation of the building and could threaten escape routes.

A lift well should be contained within the enclosure of a protected stairway or enclosed within its own fire-resisting shaft, unless it is sited so as to be neither prejudicial to an escape route nor a perforation of a compartment floor.

A passenger lift should not extend down into a basement if it is in a building with only one escape stairs or if it is within an escape stairs enclosure that terminates at ground level. If such a lift does continue to basement level it should be approached only via a protected lobby or corridor, unless it is within the enclosure of a protected stairway.

A protected lobby or protected corridor approach is also required to any passenger lift, not within a protected shaft, that serves an enclosed car park. A similar provision is required to one that serves a storey containing a kitchen, lounge or store and delivers into a storey containing sleeping accommodation.

To avoid smoke generated by a fire in the lift machinery, penetrating into an escape route, the lift machine room should be sited over the lift well where possible; if this cannot be achieved, it must be positioned outside the stairway.

The discounting of passenger lifts as a means of escape makes for difficulties in respect of the movement of disabled people in an emergency. To overcome this problem, evacuation lifts can be installed which incorporate a number of special measures that ensure the safe operation of the lift should a fire occur. Approved Document B1 does not give any details for these special measures but states that the guidance found in BS 5588: Part 8 should be followed.

This Standard states that a lift for this purpose should be either a fire-fighting lift (see 15.15.6) or a purpose-designed evacuation lift. The recommendations for an evacuation lift are:

- It should be contained within a 30 minute fire-resisting protected lift well.
- It should be approached through fire-resisting lobbies at each storey.
- There should be a protected route from the final lift exit level to the final exit from the building.
- Unless the lift serves two storeys only, there should be an evacuation lift switch adjacent to the lift landing door at the final lift exit level. The operation of this switch makes the lift return to the final exit storey level, disables the landing call controls (leaving the lift under the control of the car control panel only) and maintains the lift communication system – if one is fitted.
- There should be an alternative power supply, as described for fire-fighting lifts. The associated cables should be separated from the primary supply and routed through areas of low risk or be physically protected against breakdown.

15.12	Power circuits, emergency lighting and lighting of escape routes

Where it is necessary for power to be maintained in the event of a fire to ensure that essential equipment continues to function, protected circuits must be installed. These should be wired up with cable classified as CWZ, in accordance with BS 6378, run through those parts of the building in which the fire risk is negligible and separate from all circuits provided for any other purpose.

Emergency lighting is necessary in many of the building uses covered in this book for the maintenance of essential services, irrespective of that required for fire safety.

Escape routes and escape stairs should be provided with adequate artificial lighting powered from the mains, the lights on the corridors should be on a separate circuit from the lights on the stairs. Those routes shown in Table 15.11 should also be equipped with emergency escape lighting which will illuminate the route should the electrical mains fail.

Table 15.11   Escape lighting

Purpose group of the building or part of the building	Areas requiring escape lighting
Residential buildings	All common escape routes
Buildings where the public can assemble Recreation buildings	All escape routes and accommodation except for: (a) accommodation open on one side to view sport or entertainment during normal daylight hours (b) toilet accommodation having a gross floor area less than $8\,m^2$
Any non-residential purpose Stores	Underground or windowless accommodation Stairways in a central core Stairways serving storeys more than 20 m above ground level Internal corridors more than 30 m long
Car parks available to the public	All escape routes
Premises for any purpose	Electricity generator rooms Switch room and battery room for an emergency lighting system Emergency control room

(*Source*: Table 9 of Approved Document B1 (1992 edition))

Approved Document B1 does not define any standards for emergency lighting but points out that these are given in BS 5266: *Emergency lighting*. Part 1: *Code of practice for the emergency lighting of premises other than cinemas and certain other specified premises used for entertainment*. This British Standard will shortly be modified to include the requirements of the new EC Workplaces Directive and, since this is intended to have a retrospective application to all buildings where people are employed, new lighting installations should be designed to these standards now, where applicable. It should also be noted that this Directive will apply to *all* buildings where people are employed and thus extends the scope of the Fire Precautions Act which just applies to factories where more than 20 persons are employed.

The provisional requirements document Pr EN 1838–5 follows BS 5266: Part 1 in dividing emergency lighting into two groups: escape lighting and standby lighting. The first group ensures that escape routes are illuminated at all material times and the second enables normal activities to continue if the mains supply fails. It is the first group that is the concern of this chapter and this is further subdivided into three categories: escape route lighting; anti-panic lighting and high risk task area lighting.

Escape route lighting must have a minimum emergency duration of one hour but in certain premises where escape may not be effected quickly, duration times of up to three hours are necessary as shown below:

Hospitals, nursing homes and the like: 2 hours where there are less than 10 beds; 3 hours where there are more than 10 beds.

Premises with sleeping accommodation: 1 hour where there are fewer than 10 beds; up to 3 hours in a large hotel, etc.

Recreation buildings: 1 hour where fewer than 250 people will assemble; 2 hours if there will be more than 250 people present.

Schools, colleges and the like: 1 hour.

Sports grounds: 3 hours (Home Office *Guide to Safety at Sports Grounds*).

Luminaires must be provided to illuminate:

- escape routes to a minimum standard of 1 lux along the centreline
- important locations such as exit doors
- exit and safety signs
- staircases
- fire-fighting equipment
- first-aid points

Each compartment of the escape route must have at least two luminaires and there must also be emergency lighting in lifts, escalators, plant rooms, toilets and covered car parks.

Anti-panic lighting is to be provided to reduce confusion in the event of the mains failure and to facilitate people finding their way to the defined escape routes. The new requirement is a level of illuminance at the worst point of not less than 0.5 lux.

High risk task area lighting is introduced in the Directive for the first time. This is to highlight any places of hazard and calls for a level of emergency illumination equivalent to 10 per cent of the normal lighting level or 15 lux, whichever is the greater.

The new European Standard also places greater emphasis on the optical performance of emergency light fittings and should be taken into account when designing the lighting lay-out.

For hospitals, the recommendations of BS 5266: Part 1 are supplemented by Health Technical Memorandum 2007: *Electrical services: supply and distribution*, Health Technical Memorandum 201: *Emergency electrical services* and the Chartered Institute of Building Services Guide: *Lighting guide for hospitals and health care buildings*. To enable essential hospital services to be maintained, most building complexes are provided with standby generators designed to come into operation within 15 seconds of the failure of the electricity mains. In certain areas such as stairways or operating theatres, a faster response to mains failure is necessary and a battery backup, with a response of 0.5 second is provided.

The emergency lighting system in a hospital is divided between essential and non-essential circuits. Where these are not segregated, as is usually the case, the essential circuits should be wired in fire-resistant cable and the luminaires supplied by them should be designed to provide between 30 and 50 per cent of the normal lighting level. In addition, the luminaires in circulation spaces should be provided with their own independent circuits so that in the event of the failure of the room lights, the circulation spaces will remain illuminated and vice-versa.

Any parts of the hospital not required to be provided with the system of essential and non-essential lighting circuits as described above, should be provided with escape lighting designed as recommended in BS 5266: Part 1 with a minimum duration time of three hours.

With respect to a sports ground, the Home Office *Guide to Safety at Sports Grounds* recommends that consideration should be given to the cancellation of an event if a power failure occurs but does add that if the auxiliary power supply is capable of supplying the entire load for the ground for at least three hours it may be possible to continue with the fixture. To supply such a load a standby generator would be required rather than a backup system of batteries. In addition, if the event is to continue, a second, backup emergency power supply will be necessary.

Any auxiliary power equipment must be located in its own secure room of a one hour fire-resisting construction from which the public is excluded.

The provision of emergency lighting is covered in both the Home Office Guide and Approved Document B1. In the Home Office Guide emergency lighting is to be provided in all parts of the building to which the public has access and along all exit routes, but Approved Document B1 states that such lighting is not required where the accommodation is open to one side to view the sport during the hours of daylight (see Table 15.5). The Home Office Guide adds that the supplies to this lighting must be completely separate from the normal lighting system and, unless a separate and independent external supply is available, the emergency lighting system must connect to its own power source within the ground which can be batteries or a standby generator, with a delay of not more than 5 seconds. Whether a battery source or a generator is used it must be capable of maintaining the necessary level of illumination laid down in the IEE Lighting Code for a period of not less than three hours.

In some grounds the public address system is part of the fire-warning system, in which case it should comply with BS 5839: Part 1 and be supplied from an auxiliary power source.

15.13 Mechanical ventilation and air-conditioning systems

Any system of air movement and the ductwork associated with it is a potential source of danger in a fire due to the way it links spaces within the building. To meet this, the system should be designed either to close down if a fire occurs or to direct the air movement away from protected escape routes and exits.

In some building designs, smoke control is achieved by a pressurization system designed to propel the smoke way from escape routes. Where this has been provided, any air-handling system must be designed to be compatible with it under fire-operating conditions.

Approved Document B1 refers to two British Standards for further guidance:

- BS 5588: Part 9 for air recirculation systems operating under fire conditions and for the design and installation of ventilation and air-conditioning ductwork.
- BS 5720 for the design and installation of mechanical ventilation and air-conditioning plant.

15.14 Safety measures in air-supported structures

Air-supported structures are not, specifically, mentioned in the Building Regulations but it is recognized that they are buildings and, therefore, subject to the Regulation requirements and principles. One of the most common uses to which air-supported structures are put is to accommodate sporting activities, mainly because this is an inexpensive way of achieving a large, uncluttered enclosure. Being sufficiently light to be held up by air pressure alone, the enclosing membrane can also present a hazard in the event of a fire. Care must be taken, therefore, to ensure that the occupants can escape safely, either before the canopy collapses or from beneath it.

The following recommendations are all taken from BS 6661: *Design, construction and maintenance of single-skin air supported structures* and should represent building standards that would comply with the requirements of Regulation B1.

15.14.1 Emergency exits and routes

If the total design occupancy of the structure is less than 100, the minimum width of exit is 800 mm. Where the occupancy exceeds 100, the minimum width is increased to 1100 mm and in very large structures, used for assembly purposes, the exit width can be calculated on the basis of 530 mm per 100 persons.

The total aggregate width should be calculated using the discounting principle where one exit in turn is discounted in arriving at the widths required for the rest.

As in all buildings, there must be alternative exits at opposite ends of the structure and a sufficient number of others to ensure that no dead-end travel distance exceeds 9 m and the direct travel distance between exits is less than 30 m (see Fig. 15.7). An exception to this is in a swimming pool, where the maximum travel distance should not exceed 15 m and the access steps to the pool should be adjacent to the emergency exits.

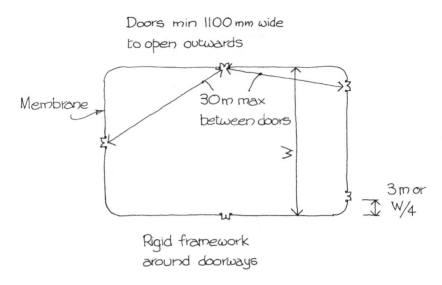

Doors min 1100 mm wide
to open outwards

Membrane

30m max
between doors

3 m or
W/4

Rigid framework
around doorways

All exit routes to have
hard well-drained surfaces

Fig. 15.7 Emergency exits in air-supported structures

Any revolving access doors should have an emergency exit located next to it.

All emergency doors should open outwards, clear of any steps, landings or other projections. They can be either single or double leaf doors fitted into an independent rigid emergency exit framework and opening either direct to the external air or into an air lock.

Exit routes must be finished with a hard, well-drained surface.

15.14.2   Emergency lighting

Emergency exit signs and the escape routes leading to the exits should be adequately lit in accordance with BS 5266: Part 1. The signs should be arranged at both high and low level so that they will not be obscured in the event of the partial collapse of the membrane.

The emergency lights should come on automatically in the event of the failure of the power source supplying the normal lighting. The emergency lighting power source should be either standby batteries capable of supplying the full load for not less than one hour, or a standby generator capable of supplying full load within 30 seconds of the failure of the main supply.

All emergency lighting wiring should entirely separate from that installed for the normal lighting. Any flexible cables should be suitably insulated with either chloroprene or butyl rubber and arranged to be out of reach. Further protection is required if the cables are liable to mechanical damage.

Electrical fittings and conductors should be adequately supported, clear of the membrane and not attached to it in any way.

15.14.3 Emergency support systems

In some air-supported structures, an emergency support system is fitted to hold up the deflated membrane, clear of any occupants. It must, therefore, be strong enough to support the mass of the membrane under short load conditions without damage. It must also maintain a headroom of at least 2.5 m over all the emergency exit routes and over 66 per cent of the total plan area.

The emergency lighting fittings should be mounted on the emergency support framework and so arranged that they illuminate the escape routes. Any hot components of the emergency lights should not be allowed to make contact with the deflated membrane.

15.14.4 Warning systems

The minimum safeguard is the provision of a low-pressure alarm and a main fan failure alarm. If the structure is to be used for public assembly, a wind speed alarm should also be provided. All these alarms should be designed to alert both the staff responsible for safety and the occupants of the building.

Consideration should also be given to the provision of fire alarms, smoke detectors linked to the fire alarms and an alarm to indicate a change-over to the emergency lighting system.

All alarms should be entirely automatic and designed and installed in accordance with the recommendations of BS 5839: Part 1.

15.15 Access and facilities for the fire services

Section 15 of Approved Document B1 gives guidance for the selection and design of facilities for the purpose of protecting life by assisting the fire service. The facilities covered are: access for fire appliances, access for fire-fighting personnel, the provision of fire mains within the building and the venting of smoke and heat from basements.

In some buildings, access for fire-fighting purposes may well be achieved by a combination of use of escape routes and externally by ladders and appliances on the perimeter. In deep basements and tall structures, fire fighters will, invariably, work inside the building and will require special access facilities and fire mains. Fire appliances will need to be able to get to entry points near the fire mains.

15.15.1   Vehicle access

Access is required to enable high reach appliances, such as turn-table ladders and hydraulic platforms, to be used and for pumping appliances to supply water and equipment for fire fighting and rescue.

Buildings up to 2000 m² in area and with a top storey less than 9 m above ground level that are without a fire main, should be designed to allow vehicular access to within 45 m of any point of the area of the projected plan, or to 15 per cent of the perimeter of the building where there is a door not less than 750 mm wide giving access to the interior of the building, whichever is the easier to achieve. Buildings over this size, not fitted with fire mains, should provide access for fire service vehicles as set out in Table 15.12. The plan area to be taken includes the outline of any overhanging upper storeys projected down onto the ground level. The perimeter length of a building is the distance round the edges of the plan area, not including the length of any walls that are in common with another building.

There should be access for a pumping appliance to within 18 m of the connection point of any dry fire mains, and this inlet should be visible from the appliance. If the building is fitted with a wet main, the pumping appliance must be able to get to within 18 m of, and within sight of, a suitable entrance giving access to the main and in sight of the inlet for the emergency replenishment of the suction tank for the main.

The access route for the fire service vehicle may be a road or other route and should meet the requirements of the fire authority and the criteria set out in Table 15.13 and Fig. 15.8. Where such access is provided, overhead obstructions such as cables or branches should be avoided within a zone

Table 15.12   Percentage of the perimeter of a building, not fitted with fire mains, which must be reachable by fire service vehicles

Type of appliance	Height of the floor of the top storey above ground level	Perimeter of the building envelope (%) which must be reachable in buildings with a total floor area (m²) of:				
		Up to 2000	2000 to 8000	8000 to 16000	16000 to 24000	over 24000
Pump	up to 9 m	See notes	15%	50%	75%	100%
High reach appliance	over 9 m	15%	50%	50%	75%	100%

(*Source*: Based on Table 19 of Approved Document B5 (1992 edition))

Notes
1. For small buildings (up to 2000 m² and top floor less than 9 m above ground level) access should be to within 45 m of any point of the plan of 15%, whichever is the less onerous.
2. There should be two site access points for fire service vehicles to health care premises. (Health Technical Memorandum 81.)
3. The height of storage buildings should be measured to mean roof level.

Table 15.13 Typical fire service vehicle access route specifications

Type of appliance	Minimum carrying capacity (tonnes)	Minimum dimensions (m) of:				
		Road between kerbs	Gateways	Turning circle between kerbs	Turning circle between walls	Clearance height
Pump	12.5	3.7	3.1	16.8	19.2	3.7
High reach	17.0	3.7	3.1	26.0	29.0	4.0

(*Source*: Based on Table 20 of Approved Document B5 (1992 edition))

Notes
1. Some fire authorities have vehicles of greater weight or other sizes and the Building Control Authority may require a different specification.
2. High reach vehicles have their load distributed over sufficient axles to allow the roadbase to be designed to take the 12.5 tonnes loading only, but any bridges should be designed to take the full 17.0 tonnes.

12.2 m out from the building and of a height equal to the height to the ridge. Any dead-end access roads should be provided with a turning circle based on the turning circle dimensions given in Table 15.13 and should not involve the fire appliance having to reverse more than 18 m.

Where the access route is a covered or enclosed roadway there should also be provided:

- ventilation adequate to remove the exhaust fumes from pumping appliances in operation
- fire telephones wired with CWZ cable (to BS 6387) and with a 24 hour battery backup
- a 3 hour maintained emergency lighting system

15.15.2 Access to the building for fire fighters

In low-rise buildings, the fire-fighting personnel will be able to gain access along the corridors provided as means of escape as well as the measures set out for vehicle access that will facilitate ladder access to the upper levels. High-rise buildings present a different problem and fire-fighting shafts must be provided (except in the case of certain hospitals) so that internal access can be obtained. Hospitals planned with hospital streets do not need fire-fighting shafts, according to Health Technical Memorandum 81. Instead there should be a fire main outlet by each department entrance positioned so that no part of the building is more than 60 m from the connection. If the hospital is of more than 5 storeys, or has a basement floor more than 10 m below ground, lifts for use by the fire brigade should be installed within the street and near the stairways. Access to these lifts should be direct from the

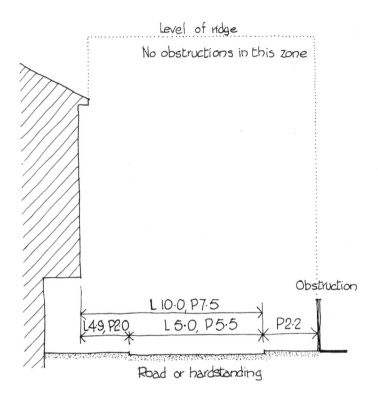

Level of ridge

No obstructions in this zone

Obstruction

L 10·0, P7·5

L4·9, P2·0   L 5·0, P5·5   P2·2

Road or hardstanding

L   Size required (in metres) for a turntable ladder

P   Size required (in metres) for an hydraulic platform

Fig. 15.8 Clearance required for high reach fire appliances

street and within 18 m of an entrance suitable for use by the fire fighters. Hospitals without streets should follow the requirements of the Regulations.

Approved Document B1 gives some guidance on the provision of fire-fighting shafts and lifts but also refers to the much more comprehensive BS 5588, particularly Part 5. The following notes are based on both the Approved Document and the main points of the Standard.

A fire-fighting shaft is a protected shaft, constructed as described for protected shafts for stairways, containing a fire-fighting lift, a fire-fighting stairs and a fire-fighting lobby. In most cases the stairs will also serve as a means of escape for occupants and for normal circulation. It is not recommended that the fire-fighting lift is considered as a means of escape, but, subject to the provision of satisfactory fire procedures and management control, it can be used for the evacuation of disabled people.

A fire-fighting shaft should not contain:

- goods lifts or service lifts
- passenger lifts, unless:
  - the access is only from a fire-fighting lobby
  - the frame and main structure is constructed of materials with limited combustibility
  - the walls and ceiling linings of the lift car have a Class 1 fire spread rating
  - the flooring of the lift car is to the same standard as that required in the fire-fighting shaft
  - the lift is clearly and conspicuously marked 'Do not use for goods or refuse'
- sanitary accommodation used as a cloakroom
- sanitary accommodation containing portable heating apparatus or any gas appliances other than a water heater or incinerator
- any cupboards
- any services except those associated with the shaft

Fire-fighting shafts are required in the following buildings:

- where there is a floor more than 20 m above ground or service access level (five storeys in the case of hospitals)
- where there is a basement more than 10 m below ground or access level
- where the floor exceeds 600 m^2 and is more than 7.5 m above ground or access level (three storeys in the case of hospitals)
- where there are two or more basement floors each exceeding 900 m^2 (1000 m^2 in the case of hospitals)

All references to hospitals above only relate to those designed without hospital street compartments.

In the first two buildings listed above, the fire-fighting shaft should contain a fire-fighting lift.

Where fire-fighting shafts are required, the number depends on whether the building is fitted with a sprinkler system complying with BS 5306: Part 2. If it is, the number of fire-fighting shafts, generally, should be:

- where the qualifying floor area is less than 900 m^2: one
- where the qualifying floor area is 900 to 2000 m^2: two
- where the qualifying floor area is over 2000 m^2: two plus one per every additional 1500 m^2 of area

If the building is not equipped with sprinklers, fire-fighting shafts should be installed at the rate of one per every 900 m^2 of qualifying floor area (or part thereof).

Hospitals requiring fire-fighting shafts should follow these same standards except that the floor areas of $900\,m^2$, given above, should be increased to $1000\,m^2$ in both cases.

A fire-fighting shaft does not need to extend down into a basement unless it is required because of either the depth or the area of the basement floors. Similarly, a fire-fighting shaft required for a basement does not have to extend up into the building unless the upper storeys require it by reason of their height or area.

Each fire-fighting shaft must connect with all the intermediate storeys between the highest and the lowest it serves and be located so that no part of the storey is more than 60 m from the entrance to a fire-fighting lobby, measured on a route suitable for laying hoses. If the internal layout is not known at the planning stage, a direct line of 40 m may be substituted.

Access to the shafts must be from the open air or via protected corridors not more than 18 m long. Where these corridors are also a means of escape they must be 500 mm wider than is required for escape purposes. Any accommodation connecting with any corridor must do so through a protected lobby.

The fire-fighting lobby at the access point should contain $5\,m^2$ of floor area clear of any escape routes so that it can act as a fire service mustering point.

Fire-fighting shafts should also be equipped with smoke control as detailed in 15.15.8.

15.15.3 Design and construction of fire-fighting shafts

Any external walls enclosing a fire-fighting shaft only have to achieve a fire-resisting standard if the shaft is designed as an independent structure connected to the accommodation and within 5 m of it, in which case either the walls of the shaft or the walls of the accommodation facing each other must possess a two hour fire-resistance rating (see Fig. 15.9). There is an exception to this rule and that is where there is a window in the fire-fighting shaft less than 500 mm from the internal angle, in which case the external wall adjacent to it must be of a one hour standard for a distance of 500 mm, also shown in Fig. 15.9.

Internal walls separating the fire-fighting shaft from the accommodation must possess a fire resistance of two hours when subjected to fire from the accommodation side and one hour from the shaft side. Walls within the shaft should be of one hour's resistance. The internal walls should also be built either of brick or concrete or should comply with the requirements of Appendix A to BS 5588: Part 5 with respect to resistance to damage. The fire resistance, durability and resistance to damage should not be significantly reduced by the effects of water from fire fighting or the operation of sprinklers.

Stairs, landings and ceilings should be built of materials with limited combustibility and all internal surfaces should possess a spread of flame rating of Class O.

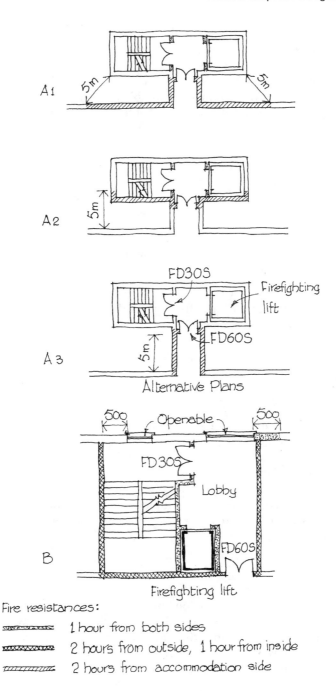

Fig. 15.9 Protection of fire-fighting shaft from external fire

A fire door should have at least half the resistance time of that required for the wall in which it is fitted, subject to a minimum resistance of 30 minutes.

Flooring should minimize slipperiness when wet; either not ignite or burning limited to a 35 mm circle from the point of burning; be firmly secured to the floor with a non-water soluble adhesive and be interrupted at all door openings with a metal threshold at least 50 mm wide.

15.15.4   Fire-fighting lobbies

The purpose of a fire-fighting lobby is to reserve, for the fire service, a protected area at each storey level in which to prepare to fight the fire. It should not, therefore, be part of the general circulation of the building and should be clearly and conspicuously marked:

'Fire-fighting lobby: do not use for storage'

In size, the lobby should have a clear floor area of not less than $5 \, m^2$ but it does not need to exceed $20 \, m^2$ if the lobby serves up to four lifts or $5 \, m^2$ per lift for lobbies serving over four lifts. None of the principal dimensions should be less than 1.5 m, nor more than 8 m if the lobby serves up to four lifts or 2 m per lift over four.

All lobbies should be equipped with smoke control as detailed in 15.14.8.

15.15.5   Fire-fighting stairs

As well as the standards set for stairs on escape routes, fire-fighting stairs have a minimum width of 1100 mm clear for a height of not less than 2000 mm except for strings protruding not more than 30 mm or handrails protruding not more than 100 mm.

A fire-fighting stairs serving floors both above and below ground level should be separated into two at ground level by a fire door.

The enclosure to the fire-fighting stairs should be equipped with smoke control as detailed in 15.15.8.

15.15.6   Fire-fighting lifts

The whole of the lift construction must be carried out with materials of limited combustion, the walls and ceiling of the lift car must have a spread of flame rating of Class 1 and the flooring should minimize the risk of slipperiness when wet. It should be firmly stuck down with a non-water solvent adhesive and either not ignite at all or only burn to a circle of not more than 35 mm.

The speed of the lift must be such as will ensure that it will run its full travel in less than one minute.

Sprinklers should not be fitted in the lift well, to avoid damage to the electrical equipment and any in the fire-fighting lobby should be sited so that they do not drench the lift landing doors or controls.

Water from fire-fighting operations must also be prevented from entering the lift well and any pieces of electrical equipment within 1 m of the wall separating the lift well from the fire-fighting lobbies (mainly the car and landing door controls), must be enclosed to protect them from dripping water. There should be no electrical equipment within 1 m of the bottom of the well.

The machine room should be within the fire-fighting shaft but not directly below the fire-fighting lift well and either:

- separated from the fire-fighting stairs, the fire-fighting lobbies and the lift well by a 1 hour fire-resisting construction, except for minimum size holes for ropes and cables; the construction must also withstand impact damage and the effects of water and the machine room must be accessible only by way of a fire-fighting lobby; or
- located above the lift well and accessible only via the fire-fighting stairs; the door between the stairs or lobby and the machine room is to be a fire door.

The lift car should be fitted with an emergency trap door and be clearly marked:

'Fire-fighting lift: do not use for goods or refuse'

unless it is the only lift in which case it should be marked:

**'Fire-fighting lift: do not obstruct the doors.
Do not leave goods in the lift'**

At the fire service access level the fire-fighting lift must be clearly marked:

'Fire-fighting lift'

Fire-fighting lift controls receive a lot of attention in BS 5588: Part 5, much of which is in the domain of the lift specialist but the following is a summary of the main points.

The whole system should be changed over to the fire-fighting mode by a fire-fighting lift switch located near the lift door at the fire service access level, protected by a locked cover opened by a standard lift door emergency key. The effect of operating this switch is to disable the normal lift controls and landing call buttons and enable the fire service lift control panel. All lifts should return to the fire service access level when the doors should open to discharge any occupants and then close. All lift car doors except that to the fire-fighting lift must remain closed to prevent the lift well acting as a chimney. There should be audible and visual warnings that the fire-fighting lift switch has been operated, at all points within the lift well, motor room and any other area entered for maintenance purposes.

The operating of the fire-fighting lift switch should also activate the fire service communication system. This should provide for two-way speech communication between the lift car, the fire service access level and the machine room. The equipment in the lift car should be a built-in microphone and speaker, not a telephone handset.

The controls within the lift car must have, in addition to the normal markings:

'Fire service access level' or 'FSAL'

adjacent to the appropriate control and indicator.

After fitting and testing a fire-fighting lift, the lift installer should issue a certificate that states that the tests and checks listed in Appendix C to BS 5588: Part 5 have been satisfactorily completed.

15.15.7   Fire mains

The purpose of fire mains is to provide the fire fighters with points to which they can connect their hoses to fight a fire from within the building. They may be 'dry mains', in which case they are empty pipes which are connected at ground level to a hose from the fire service pumping appliance, or they may be 'wet mains' which are kept full of water supplied from tanks and pumps in the building. Wet mains should be provided with the facility to replenish the supply from a fire service pumping appliance.

Any building less than 60 m high can be fitted with either wet or dry mains; over this height it should always be a wet rising main.

Further guidance can be obtained from Sections 2 and 3 of BS 5306: Part 1 on the design and construction of fire mains.

Fire mains, generally, are only required in buildings that have fire-fighting shafts. They are also required in hospitals with hospital street compartments, fitted adjacent to each department entrance so that all parts of the floor area are within 60 m of the fire main.

The valved outlets from the mains should be located in each fire-fighting lobby giving access from the fire-fighting shaft to the accommodation and at fire service level. The valves and outlets should be sited and directed so that:

- access to them is unrestricted
- they point away from lift doors to minimize the risk of water flowing into the lift well or coming into contact with lift controls and communication equipment
- hoses can be connected, charged and advanced into the accommodation without excessive kinking or the obstruction of doors and exit routes

15.15.8   Electrical services in fire-fighting shafts

It is essential to maintain the electrical services within the fire-fighting shaft in the event of a fire interrupting the normal supply. A secondary power supply, such as an automatically started generator or supply from a second substation, independent of the primary supply, must be provided, with sufficient capacity to permit the automatic recall to fire service access level of all other lifts in the fire-fighting shaft and to maintain in operation:

- the fire-fighting lift
- normal lighting within the fire-fighting shaft
- the fire service communications system
- any smoke control mechanical ventilation or pressurizing systems
- any pumps required to feed the fire service main

The secondary supply must come into operation within 30 seconds of the failure of the primary supply and, if it is from a generator, should be capable of providing the necessary power for three hours without replenishment of fuel.

Cables supplying current to the fire-fighting shaft should be installed in accordance with the IEE Wiring Regulations. The cables should be located in a protected shaft, where possible in the lift well or adequately protected from fire, for a period not less than that required for the structural protection of the shaft, or classified as CWZ in accordance with BS 6387.

Cables, other than those required for the operation of the fire-fighting lift, should be located outside the lift well but inside the fire-fighting shaft.

Both the primary and the secondary power supply cables should terminate in an automatic change-over device located within the fire-fighting shaft that will effect a transition from primary to secondary supply if any phase of the primary power system fails. All switchgear controlling supplies to the fire-fighting shaft should be clearly labelled:

'Fire-fighting shaft: do not switch off'

An indication of the status of:

- the primary and secondary power supplies
- any mechanical ventilation or pressurization systems
- any pumps feeding fire mains

should be provided adjacent to the fire-fighting lift switch and duplicated in the fire control room (if provided).

The fire-fighting stairs and lobbies should be provided with an emergency lighting system complying with BS 5266: Part 1.

15.15.9 Smoke control

Smoke is one of the greatest problems faced by the fire fighters and should be prevented from entering, or be swiftly removed from, the fire-fighting shafts, fire-fighting lift enclosures or fire-fighting lobbies.

Two systems are detailed in BS 5588: Part 5: pressurization and natural ventilation.

All shafts serving basements more than 9 m below ground must be equipped with a pressurization system; all other shafts can have either a pressurization system or use natural ventilation.

The pressurization system must be designed in accordance with BS 5588: Part 4 and activated by smoke detectors mounted at high level in the accommodation adjacent to the doors leading to the fire-fighting lobbies.

Natural ventilation should consist of an openable vent equal to 5 per cent of the stairs enclosure area fitted at the end of the enclosure and operated by remote control from a point adjacent to the fire service access door. This operating point must be clearly marked as to its function and operation.

Where the stairs are adjacent to an external wall, openable vents can be fitted, equivalent to 15 per cent of the stairs enclosure area at each storey level above the ground level.

Any stairs that serves only a basement less than 9 m deep and leads directly to the final exit does not need to be fitted with openable vents.

Fire-fighting lobbies at ground level should have openable vents equal to 25 per cent of the floor area and at each basement level, vents of $1\,m^2$ at high level discharging either into the open air or (more probably) into a smoke shaft not less in size than the vent. These smoke shafts must discharge directly to the open air at ground level at a point where neither the exits from the building nor the fire access will be affected.

Smoke vents should be:

- clearly identifiable and accessible
- constructed to open out to a minimum angle of 30°
- fitted with:
  - simple lever handles, or
  - rotary drives to simple rack or gear operated devices, or
  - locks operated by a square-ended key, 8 mm × 8 mm in section and 25 mm deep

Permanent ventilators should not be provided for these purposes.

15.15.10   Venting of basements

The products of combustion tend to escape via the stairways that are to be used by the fire-fighting personnel, making access, search, rescue and fire fighting difficult. The provision of smoke outlets – also called smoke vents – allows the smoke and heat to escape and can also be used to allow cooler air into the basement.

Where practicable smoke outlets should be provided in each basement space. It may be that the building shape is such that there are basement areas that are not on the perimeter. In these situations it is acceptable that only the perimeter spaces are vented directly, the inner areas being vented indirectly by the fire fighters opening the connecting doors. If, however, the basement is compartmented, each compartment must be directly vented and not have to rely on opening doors in the compartment walls.

Smoke outlets are not required in a basement where the floor area is less than $200\,m^2$ and not more than $3\,m$ below ground level, nor where the basement is a strong room.

Naturally venting outlets should be sited either in the ceiling or at high level in the wall of the basement, evenly distributed around the perimeter of the building so as to discharge to the open air. Their total free area should be at least 2.5 per cent of the floor area of the space they serve and separate outlets should be provided at places where there is a special fire risk.

Generally the outlet can be covered by a panel, stallboard or pavement light that should be clearly indicated and which can easily be broken out. If the outlet has to be at a point that is not easily accessible it should only be covered by a non-combustible grille or louvre. The position of the smoke outlets must be carefully selected to avoid causing any difficulty in the use of a fire escape route from the building.

Basements fitted with an automatic sprinkler system that conforms to the recommendations of BS 5306: Part 2 can also have a mechanical smoke extract system that is activated by the sprinkler system. Alternatively, it may be activated by an automatic fire detection system conforming to BS 5389: Part 1 (at least L3 standard). The mechanical extract system should be capable of achieving at least 10 air changes per hour and handling gases at temperatures up to 400 °C for not less than one hour.

The outlet ducts and shafts should be enclosed by a non-combustible construction.

15.16 Alternative approach

The provision of means of escape is comprehensively covered in BS 5588: Part 1 in relation to residential buildings, BS 5588: Part 6 in relation to places of assembly and BS 5588: Part 8 in relation to the means of escape for disabled people. Approved Document B1 guidance corresponds in most respects to the recommendations contained in these British Standards and states that some must be observed but the Document points out that if the British Standard is adopted as an alternative to the Approved Document, all the relevant provisions of the Code should be followed, rather than a mixture of the Standard and the provisions of the Approved Document.

Chapter 16

PROVISIONS FOR DISABLED PEOPLE

16.1 The Building Regulations applicable

The Building Regulations concerned with access and facilities for disabled people are all contained in Part M and the associated Approved Documents M1 to M4 but cross-reference is made in other Approved Documents, particularly B1 on fire safety and K on stairs and ramps. The specific requirements are:

M1 Interpretation
M2 Access and use
M3 Sanitary conveniences
M4 Audience and spectator seating

M1 defines what is meant by disabled people; M2 is concerned with the ability with which the disabled can gain access to, and the ease with which they can move about within, a building; M3 stipulates that if sanitary conveniences are provided in the building, reasonable provision should be made for people with disabilities; M4 requires that if a building contains audience or spectator seating, reasonable provision shall be made to accommodate disabled people.

In addition to the Building Regulation requirements, there is also the Chronically Sick and Disabled Persons Act 1970, amended in 1976, which states that buildings are to be designed so as to be accessible to disabled people wherever reasonable and practicable. This legislation would, no doubt, be adequately met by compliance with the constructions and standards set out in Approved Document M.

Further specific guidance is given in Design Note 18, published by the Department for Education with the title *Access for disabled people to educational buildings*. Where relevant, any details that differ from those given in the Approved Documents are quoted in this chapter.

16.1.1 The application of the Regulations

The term 'disabled people' in this context is taken to mean all people who have any impairment to their mobility that either limits their ability to walk, or requires them to use a wheelchair, or who have an impairment of their hearing or of their sight.

The provisions stipulated are for the benefit of both the people who normally occupy the building or who are visitors to it.

The requirements of Approved Document M apply to a new building, to the reconstruction of an existing building and to an extension to an existing building. In the case of a reconstruction, if it is impractical to adjust the existing entrance to permit independent access by wheelchair users, or to create a new and suitable entrance, the other requirements of Part M should still apply. Where a building is extended there is no obligation to alter the existing to bring it up to the standard of these requirements but the extension should not adversely affect the existing building with respect to these provisions. If the extension is approached through the existing building, the Approved Document states that it would be unreasonable to require higher standards than that existing, but where the extension is independently approached and entered, it should be treated as a new building.

In addition to provisions within the building, Part M also deals with those external features that are needed to provide access from the boundary of the site or a car park to the building.

16.1.2 Objectives

The purpose of the legislation as applied to any place that people assemble or are in residence is to ensure that:

- Those who are disabled can reach the principal and other entrances to the building from the boundary or a car park.
- Elements of the building do not create a hazard for a person with impaired vision.
- Disabled people can gain access into and within any storey of the building and to any facilities provided for them.
- Disabled people can use the facilities and services the building offers.
- Any meeting rooms or large reception areas in excess of $100\,m^2$, booking areas and ticket offices are equipped with aids to communication so that a person using a hearing aid is afforded the benefit of receiving sound without loss or distortion through bad acoustics or extraneous noise.
- Any disabled person can be accommodated within an audience or spectator seating area at a standard equivalent to that provided generally.
- There is suitable sanitary accommodation for disabled people.

The requirements for aids to communication can be fulfilled by a loop induction system or an infra-red system. Both suffer from disadvantages in use; the former can compromise confidentiality and the latter requires the listener to wear a stethoscope for reception. It is left to the building owner to decide the most suitable system.

16.2   Tactile paving

One of the ways of meeting the problems encountered by someone with defective vision is to provide advance warning of any changes in the paving of the approach to the building. This can be achieved by tactile paving slabs that have a surface modelling that can be detected by someone stepping on it. Two types are shown in Approved Document M2, corduroy and blister paving. The first of these has parallel ribs running across its face and the second has a pattern of bumps. Both are illustrated in Fig. 16.1. For additional assistance the paving can be in a contrasting colour, although the Approved Document does not mention this.

Blister paving should be laid wherever a route crosses a carriageway and corduroy paving should be laid at the top of a flight of steps.

Tactile paving is not a requirement inside the building.

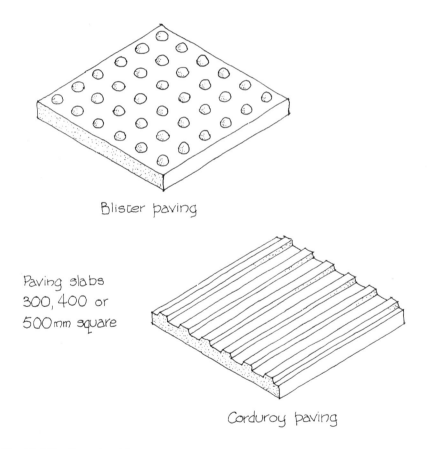

Blister paving

Paving slabs
300, 400 or
500mm square

Corduroy paving

Fig. 16.1 Tactile paving slabs

16.3   External access to and into the building

This is the first of the objectives listed above and requires the consideration of a number of points. It should be remembered that the needs of disabled people vary and alternatives are helpful, for instance, a person with walking difficulties but not needing to use a wheelchair may well find steps easier to negotiate than a ramp. It is also important to avoid any abrupt changes of level that can present a hazard to wheelchair users or to someone with impaired sight. Where at all possible the gradient should not be steeper than 1 in 20 (Approved Document M2 refers to this as a 'level' approach). If this cannot be achieved, the rules applying to a ramped approach must be followed.

It is also necessary to allow sufficient width for someone using a wheelchair, sticks or crutches, or who is blind or partially blind plus another person accompanying them to proceed in one direction, leaving room for another person to pass the other way. For this reason the minimum width of a path, intended for use by the disabled, that would satisfy the requirement is 1200 mm. This same path should also be provided with tactile warning surfaces (corduroy paving) at the top of any steps and a similar surface (blister paving) plus dropped kerbs if it crosses a carriageway.

Another hazard for people with limited sight approaching a building is outward opening windows and doors. In any position where a path runs parallel to the external wall of the building, it should be kept away from the face of the wall by a distance clear of the outward swing of a window or door and a very slight change of level and strongly tactile paving, like cobbles, laid between the path and the wall. Alternatively, a guard rail or similar provision should be fitted to protect a disabled person.

Since the aim is to provide access for the staff working in the building and for members of the public (if it is intended that the latter should visit the building), suitable approaches must be constructed to the principal entrance and to any entrances intended for the exclusive use of staff.

16.3.1   Ramped approach

As already mentioned, where a 'level' approach cannot be achieved, a ramped approach must be constructed. A long ramp must be less steep than a short one but neither should exceed a safe gradient. Below 1 in 20 the approach is considered to be level, ramps from 5 m to 10 m long should be graded between 1 in 20 and 1 in 15 but slopes less than 5 m long can be graded between 1 in 15 and 1 in 12. The requirements for educational buildings are similar and Design Note 18 states that for a ramp not exceeding 3 m long, the maximum gradient is 1 in 12, from 3 to 6 m long it is 1 in 16 and from 6 to 10 m it is 1 in 20. In all cases the length should not exceed 10 m.

The needs of disabled people and their helpers that should be considered are:

• Frequent landings to enable them to rest, regain breath or ease pain.

- Adequate space on landings to allow room to manoeuvre a wheelchair.
- Supports on both sides of the ramp as a disabled person may have weakness in their left or in their right side.
- Kerbs or solid balustrades on the open side, or sides of the ramp to avoid the risk of wheelchair users catching their feet beneath or between balustrade rails or standards.

The details of the dimensions that should be followed in the construction of a ramp are given in Chapter 10 at Section 10.9 but are reproduced below and illustrated in Fig. 16.2:

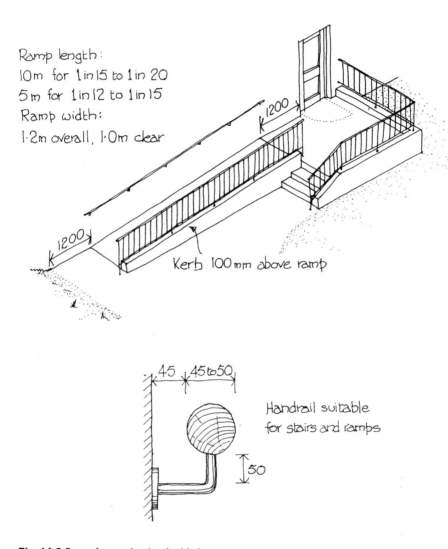

Ramp length:
10m for 1 in 15 to 1 in 20
5m for 1 in 12 to 1 in 15
Ramp width:
1·2m overall, 1·0m clear

1200

1200

Kerb 100mm above ramp

45   45 to 50

Handrail suitable for stairs and ramps

50

Fig. 16.2 Ramp for use by the disabled

- Width: surface width, 1200 mm, unobstructed width, 1000 mm; a surface width of 1500 mm is preferred in an educational building.
- Landings: top and bottom landings, 1200 mm long, intermediate landings, 1.5 m long, all clear of any door swings; in an educational building the landing length should be 1500 mm.
- Handrail: both sides of any ramp more than 2 m long and any landings, 900 mm above the surface of the ramp, 1000 mm above the surface of any landings. If the building is a school to be used by younger children, Design Note 18 states that an additional handrail at an intermediate height of 600 mm should be fitted. The handrail should terminate in a closed end which does not project into the route of travel and of the profile shown in Fig. 16.2.
- Kerb: 100 m high to any open edge of ramp or landing.
- Surface: ramp surface finish to reduce the risk of slipping (there is no requirement in the Approved Document for the provision of tactile paving at the head of a ramp although this would be consistent with the philosophy contained in the Document).
- Where practicable, easy going steps should be provided alongside the ramp.

Design Note 18 also required that ramps, landings and passing bays must be adequately lit.

16.3.2 External stepped approach

While the same considerations apply to steps as apply to ramps, there are some additional ones:

- Width: 1000 mm minimum, unobstructed.
- Maximum rise: 150 mm.
- Minimum going: 280 mm (measured at 270 mm from the inner edge of a tapered tread).
- Maximum total rise between landings: 1200 mm.
- Risers not to be of the open type.
- Step nosing to be distinguishable by a contrasting brightness.
- Steps should be of a suitable profile as shown in Fig. 16.3.
- Top, bottom and, if necessary, intermediate landings: 1200 mm long, clear of door swing onto it.
- Top landing to have tactile panel (corduroy paving) at least 1500 mm wide and 800 mm long, not more than 400 mm from the top step.
- Handrail 900 mm above the pitch line to both sides of any flight of more than two risers and 1000 mm high to both sides of all landings. The handrail should extend at least 300 mm beyond the top and bottom nosings and terminate in a closed end that does not project into the route of travel and be of a suitable profile as shown in Fig. 16.2.

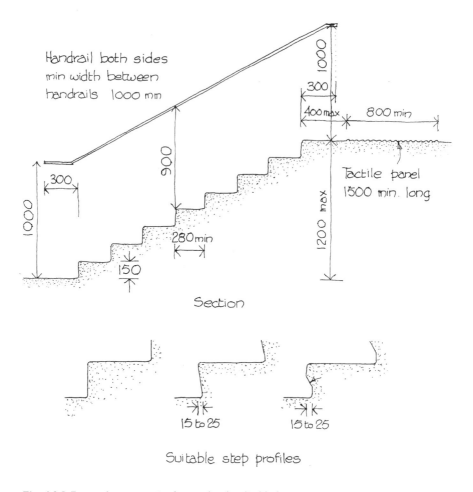

Handrail both sides
min width between
handrails 1000 mm

900

300

1000

150

280min

1000

300

400max

800 min

1200 max

Tactile panel
1500 min. long

Section

15 to 25

15 to 25

Suitable step profiles

Fig. 16.3 External access stairs for use by the disabled

These details are also covered in Chapter 10 at Section 10.4 and illustrated in Fig. 10.2, which, for convenience, is reproduced here as Fig. 16.3.

16.3.3 Access into the building

Having negotiated the approach to the building, the disabled person must then gain access into it. For this purpose, the principal entrance must be suitable whether the person is on foot or in a wheelchair. If, due to site constraints, it is not possible to make the principal entrance accessible to the disabled, a second and suitable entrance must be formed, available to all, which is connected internally to the principal entrance. Similarly a second and suitable entrance must be provided if the principal entrance is not accessible, by the disabled, from the car-parking spaces serving the building.

Sufficient clearance must be provided adjacent to and through the doorway for any necessary manoeuvres by wheelchair users and a clear view through the door made available to avoid any collisions. The details of provisions that would satisfy the requirements are:

- An opening width of 800 mm minimum, clear of the door stops and door thickness.
- An unobstructed space of at least 300 mm adjacent to the opening edge of the door, unless the door is opened by automatic means.
- A glazed panel in the door leaf from at least 900 mm to at least 1500 mm above the floor level.

It should be noted that the minimum door width applies to each leaf if a double door is fitted, which means, in practice, that, for a single door, a door set of 1000 mm is required and for a double door it would have to be 1800 mm wide.

Large revolving doors are considered suitable as they usually revolve very slowly and allow space for a wheelchair plus another person, but small revolving doors would only be permissible if there is also a hinged door adjacent complying to the details listed above. This may also be necessary for fire escape purposes (see Chapter 15).

If an entrance lobby is provided inside the entrance door, its size should be such that a wheelchair-borne person can move clear of the first door before using the second. The sizes necessary to achieve this are shown in Fig. 16.4.

16.4   Movement within the building

Once within the confines of the building and its smaller spaces, the ambulant disabled person usually encounters less trouble than somebody in a wheelchair. As a result, the space provisions detailed in Approved Document M2 relate more to the needs of this latter group than to the former although the needs of those with impaired sight or hearing are not ignored.

16.4.1   Corridors and lobbies

The width of a corridor depends on whether a wheelchair user has access to it and, if so, is intended to provide sufficient room for such use and for another person to pass. For this purpose the minimum unobstructed width is 1200 mm, but if the corridor is only accessible by a stairway then a minimum width of 1000 mm is considered adequate.

Internal lobbies are less likely than entrance lobbies to be in demand by several people at the same time and, therefore, slightly smaller dimensions can be used in design, as shown in Fig. 16.4

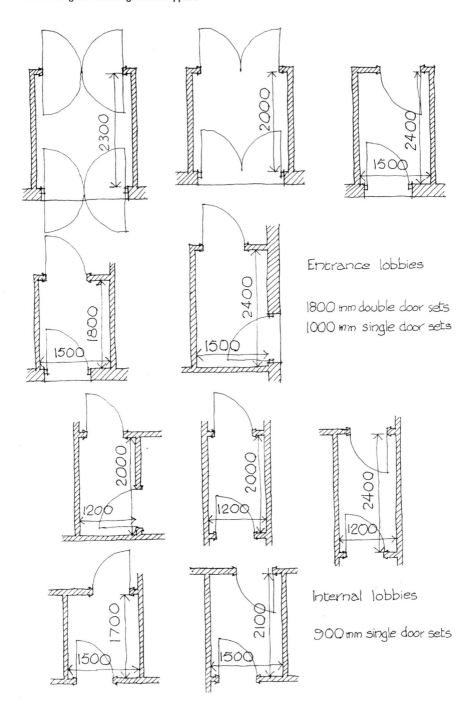

Fig. 16.4 Minimum lobby sizes

16.4.2   Internal doors

Any door intended to be used by a person in a wheelchair must be wide enough for the chair to pass through and have sufficient space at the side to allow the disabled person to move out of the door's way as they open it. Approved Document M2 sets out the minimum critical dimensions as (see Fig 16.5):

- a clear opening width of at least 750 mm
- an unobstructed space adjoining the opening edge of the door of at least

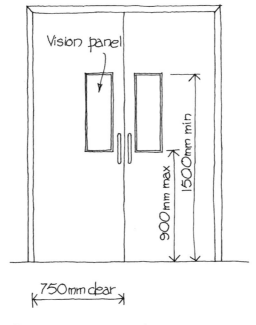

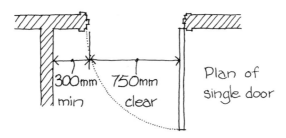

Fig. 16.5 Internal doors to be used by disabled people

300 mm, unless the door can be opened by automatic control or where it is reasonable to expect that assistance will be available such as when leaving the room of a fellow able-bodied resident

- each door across a corridor to which the disabled have access is provided with a glazed panel extending from not more than 900 mm from the floor up to at least 1500 mm from the floor

There are a number of additional recommendations when the door is in an educational building:

- Door handles should be between 830 and 900 mm above the floor if it is a nursery, primary or middle school and between 830 and 1040 mm above the floor in a secondary school or college.
- Kicking plates should be fitted to the bottom of the door, 200 mm high for normal use and 400 mm high where regular wheelchair use is anticipated.
- Door thresholds should be avoided.
- Door mats, if provided, should be flush with the floor surface and close fitting.

16.4.3 Stairs

It is necessary to provide a stairway that is suitable for people with walking difficulties if there is no lift in the building and all stairs should be suitable for people with impaired sight.

The requirements for stairs for the ambulant disabled are almost identical to those for a stepped approach to the building. The only difference being that, being inside, the risers can be slightly greater, the total rise can be more and the going slightly less, resulting in a steeper staircase. It is not considered necessary to provide advance warning of the stairs by the provision of tactile paving.

The details of a stairway suitable for use by people with walking difficulties are:

- Width: 1000 mm minimum, unobstructed.
- Maximum rise: 170 mm (163 mm in a school).
- Minimum going: 250 mm (measured at 270 mm from the inner edge of a tapered tread).
- Maximum total rise between landings: 1800 mm.
- Risers not to be of the open type.
- Step nosing to be distinguishable by a contrasting brightness.
- Steps should be of the same profile as external steps (see Fig. 16.3).
- Top, bottom and, if necessary, intermediate landings: 1200 mm long, clear of door swing onto it.
- Handrail 900 mm (850 in a school) above the pitch line to both sides of any flight of more than two risers and 1000 mm high to both sides of all landings. The handrail should extend at least 300 mm beyond the top and

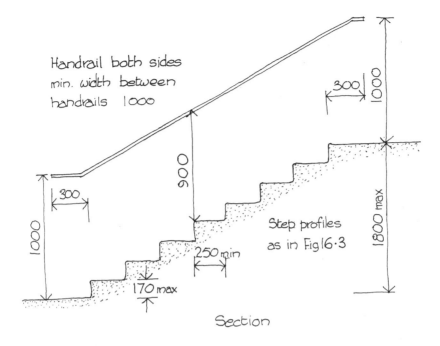

Handrail both sides
min. width between
handrails 1000

300

1000

900

300

1000

250 min

170 max

Step profiles
as in Fig 16·3

1800 max

Section

Fig. 16.6 Internal stairs for use by the disabled

bottom nosings and terminate in a closed end that does not project into the route of travel and be of a suitable profile, as shown in Fig. 16.2.

In addition, if the stairway is in an educational building, all treads and landings must be adequately illuminated at all times that the building is in occupation.

These details are also covered in Chapter 10 at Section 10.4 and illustrated in Fig. 10.3, which, for convenience, is reproduced here as Fig. 16.6.

16.4.4 Stairlifts

In some buildings there are small areas, which must be greater than $100\,\mathrm{m}^2$, with a unique function that should be accessible to wheelchair users but are not such as would justify the expense of an enclosed passenger lift. Examples of this unique function are: a staff rest room, a staff training room or a gallery in a school library.

In this circumstance the Approved Document M2 states that it is reasonable to install a wheelchair stairlift. This must be in accordance with the recommendations of BS 5776: *Specification for powered stairlifts*.

16.4.5 Ramps and platform lifts

Where there is a change of level in a storey accessible to wheelchair users, it would be reasonable to provide a ramp for their convenience. This should comply with the requirements set out under 16.3.1; however, the provision of such a ramp can create serious problems with the plan because of the lengths involved in obtaining the necessary gentle gradients. The requirements of the Regulation can be met by the provision of a platform lift if it is impracticable to provide a ramp, but there should also be a stairway that is suitable for ambulant disabled people.

If it is provided, a platform lift should comply with the recommendations contained in BS 6440: *Powered lifting platforms for use by disabled people.*

16.4.6 Passenger lifts

The most suitable means of travel from one storey to another for a disabled person is a passenger lift. However, Approved Document M2 recognizes that the cost of such an installation is such that it is unreasonable to require it in every case. The buildings where a lift is required are:

- A two-storey building with net floor areas in each storey exceeding $280\,m^2$.
- A building of more than two storeys and net floor areas exceeding $200\,m^2$.

Net floor area means the area of all parts of the storey (not including the areas of stairs, lifts, sanitary accommodation or maintenance areas) which use the same entrance from the street, whether or not they are in the same part of the storey or used for different purposes.

Where a lift is provided there must also be suitable means of access from the lift to the rest of the storey.

The requirements for the lift itself are shown in Fig. 16.7 and listed below:

- There is a clear landing of at least 1500 mm × 1500 mm in front of its entrance.
- The door opening has a clear width of 800 mm, whether it is a single door or double doors.
- The width of the car is at least 1100 mm and its depth from the entrance is not less than 1400 mm.
- The landing and car controls are between 900 and 1200 mm above the floor level of both the landing and the car, and the car controls are not less than 400 mm from the front enclosure of the car (in some installations two sets of call buttons are provided at different heights for the convenience of both able-bodied persons and those in wheelchairs, but this is not a requirement).
- A lift car installed in an educational building should have a handrail fixed to the rear and side walls at 1000 m above the car floor, or 900 mm high if the lift is for the use of young children.
- Suitable tactile indication of the floor level selection is provided adjacent to

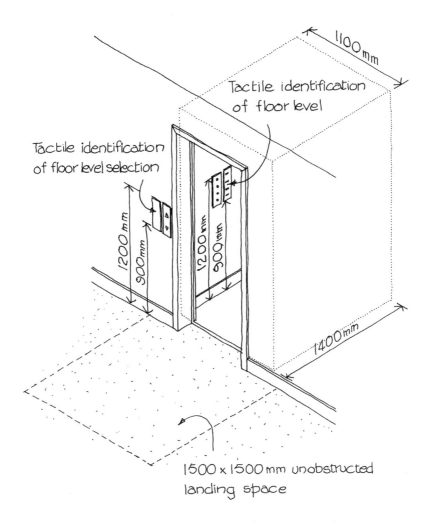

Fig. 16.7 Lift dimensions for wheelchair users

the lift call buttons on the landing and, where the lift serves more than three floors, similar tactile indicators are fitted in the lift car if it serves more than three storeys.

- Visual and voice indication of the floor reached is provided if the lift serves more than three storeys.
- There is a signal giving 5 seconds notification that the lift is answering a landing call.
- There is a 5 second 'dwell time' before the doors begin to close after they are fully open. This may be overridden by an infra-red or photo-eye door re-activating device (but not by one relying on edge pressure) provided that the minimum time for the doors to stay open is 3 seconds.

347

Details of some of these provisions are contained in BS 5655: Parts 1 and 2, Part 5: *Specification for dimensions of standard lift arrangements* and Part 7: *Specification for manual control devices, indicators and additional fittings.*

It should be noted that, although it is not a requirement of the Approved Document, BS 5655 recommends that automatic doors to passenger lifts should be equipped with re-opening activators. These may be operated by invisible beam or physical contact.

16.4.7 Hotel bedrooms

Since wheelchair-borne people will need to use the bedrooms in a hotel, consideration should be given to the amount of space they need to be able to manoeuvre themselves into and around the room. Approved Document M2 states that the requirements will be satisfied if:

- one guest bedroom in 20 (or part thereof) is designed following the dimensions given in Fig. 16.8
- the room door has a clear opening width of at least 750 mm and an unobstructed space at least 300 mm wide at the opening edge, unless the door can be opened automatically, as shown in Fig. 16.5
- the door to any other guest's bedroom has a clear opening width of 750 mm (but does not need to have the 300 mm unobstructed space)

16.4.8 Dressing cubicles and showers

All buildings where recreational activities are to be carried on should provide appropriate changing and showering facilities but, unless some of these are designed with the needs of the wheelchair-borne person in mind, many members of the public will be prevented from using the facilities offered. Because of the private nature of the function of these rooms the dimensions and equipment should be so arranged that a disabled person can attend to their own needs unless their incapacity is such that they always require assistance.

The critical issues are:

- Adequate manoeuvring space for a wheelchair.
- Adequate room and facilities to enable the user to transfer to a seat.
- Correct fixing height of seats, taps, shower heads, mirrors and clothes hooks.

The detailed recommended dimensions and lay-outs for dressing cubicles are shown in Fig. 16.9 and for showers in Fig. 16.10.

16.4.9 Restaurants and bars

Following the principle that disabled people should be able to enjoy the facilities available to all, Approved Document M2 sets out minimum standards

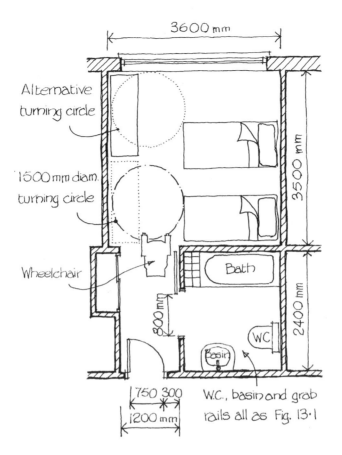

Fig. 16.8 Hotel bedroom suitable for wheelchair user

of access in restaurants and bars since it is reasonable that someone in a wheelchair would want to use these facilities on their own as much as anyone else. The Approved Document recognizes that, in some interiors, the use of steps is as much a part of the design image as a need to gain access from one level to another. For this reason the requirement for access by someone in a wheelchair is that they should be able to get to at least half of the seating areas without trouble.

Where premises offer both self-service and waiter service it should be arranged that a disabled person can gain access to, and make use of, whichever form of service they choose.

16.5 Sanitary accommodation

An important provision in a building to which the disabled have access is suitable and convenient sanitary accommodation. It should be planned and

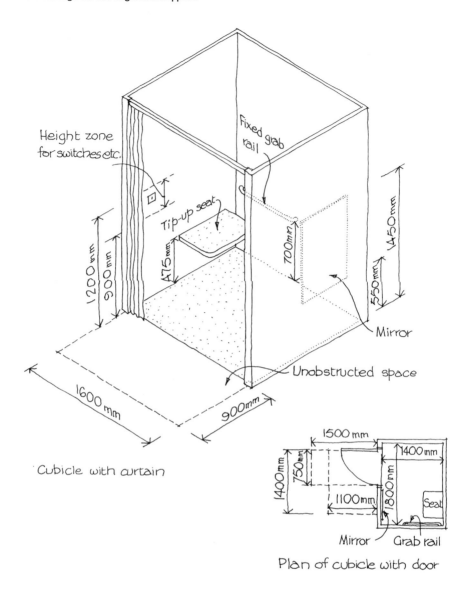

Height zone for switches etc.

Fixed grab rail

Tip-up seat

1200 mm
900 mm
475 mm
700 mm
1450 mm
550 mm

Mirror

Unobstructed space

1600 mm
900mm

Cubicle with curtain

1500 mm
1400 mm
1400mm
750 mm
1100mm
1800 mm
Seat

Mirror   Grab rail

Plan of cubicle with door

Fig. 16.9 Dressing cubicles for the disabled

equipped to be used by anyone with any form of disability and located in a suitable position, bearing in mind that a disabled person frequently needs to gain access to it quickly.

As sanitary accommodation for the disabled is almost always located with or within that provided generally, the whole of this subject has been covered in detail in Chapter 13.

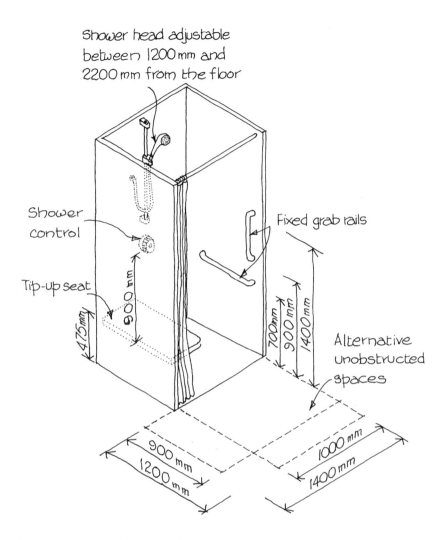

Shower head adjustable between 1200 mm and 2200 mm from the floor

Shower control

Tip-up seat

475 mm

900 mm

Fixed grab rails

700 mm
900 mm
1400 mm

Alternative unobstructed spaces

900 mm
1200 mm

1000 mm
1400 mm

Fig. 16.10 Shower cubicle for disabled person

16.6 Aids to communication

Ambulant difficulties are not the only form of disability that can cause problems for the user of a building; impaired sight and hearing also need to be taken into account. Reference has been made elsewhere to the use of tactile surfaces for the benefit of those with poor eyesight and in Approved Document M2 guidance is given on the suitability of audio systems to aid people who are deaf.

The need for this assistance occurs mostly in buildings where there is a public performance to which to listen or there is a discussion in which all should be able to play their part.

To satisfy the requirements of Regulation M2, aids to communication should be installed in auditoria and meeting rooms in excess of 100 m^2 in area, large reception areas and any ticket or booking offices where the customer is separated from the clerk by a glazed screen.

The requirement given in the Approved Document is for the person with the impaired hearing to receive a signal some 20 dB above that received by someone with normal hearing. It is also important that, whatever system is used, any reverberation and audience or other environmental noise is suppressed.

Two communication aids are reviewed: the induction loop and the infra-red system.

The first takes a signal from a microphone and passes through an amplifier, into a loop of wire round the space to be served. The magnetic field thus generated can be picked up by a listener's normal hearing aid and converted into familiar sound. An advantage of the system is that it links into the hearing aid that the disabled person customarily wears. A disadvantage is that with the loop round the perimeter of the area covered, the magnetic field can spill over into adjoining spaces and confidentiality is thus violated.

The infra-red system converts the signal into invisible light that is picked up by a personal receiver, demodulated and converted into recognizable sound. This system is less likely to suffer from the lack of confidentiality but the users must wear a stethoscope type of hearing aid to be able to receive it.

Exactly which system is installed is left to the discretion of the building owner, bearing in mind the particular function of the building and its layout.

16.7 Wheelchair spaces within an auditorium

Continuing the principle that disabled people should not be prevented from enjoying the normal facilities of a building because of their disability, Approved Document M4 sets out standards for the accommodation of wheelchair spaces within audience or spectator seating in an auditorium or sports ground.

The scale of provision is that there shall be a number of wheelchair spaces equivalent to 1/100th of the number of fixed seats with a minimum number of six. There is a slight relaxation of this rule in that in a very large stadium the requirement is rather onerous and a smaller proportion of wheelchair spaces would be reasonable. These spaces should be dispersed among the fixed seating so that a disabled person's able-bodied companion can sit beside them. In many cases two spaces are provided so that disabled people can sit together when appropriate.

The wheelchair space requirement is a clear width of at least 900 mm and a

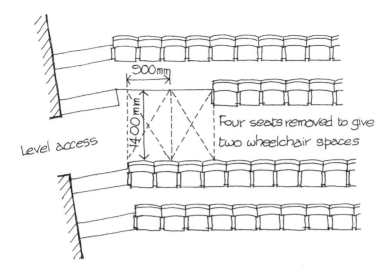

900 mm

1400 mm

Level access

Four seats removed to give
two wheelchair spaces

Fig. 16.11 Wheelchair spaces in a theatre

depth of 1400 mm from which a clear view of the event can be obtained. The space does not have to be permanent but should be easily created for the occasion by the removal of the appropriate number of seats, as shown in Fig. 16.11.

Obviously, the position of the spaces has to be one that is possible and easy to reach with a wheelchair, so must have adequate room for manoeuvre and not require the negotiation of any steps.

Chapter 17

CAR PARK BUILDINGS AND SURFACE PARKING AREAS

17.1   The Building Regulations applicable

Brief reference is made in a number of places in the Building Regulations and the Approved Documents to the measures necessary in car parks. In some cases, there is no specific mention of this application but the consequences of providing a car park must be borne in mind; for instance, a suspended floor intended to serve as an area for the parking of cars must be designed to take the appropriate load and if enclosed must have internal linings complying with the requirements for the restriction of the spread of fire. Direct reference can be found in the following:

B1   Means of escape
B3   Internal fire spread (structure)
B4   External fire spread
F1   Means of ventilation
H3   Rainwater drainage
M2   Means of access for disabled people

B1 is concerned that there is a means of escape to a place of safety outside the building, and this is particularly critical in enclosed car parks; B3 deals with the design of the structure and its ability to maintain its stability for a reasonable period in the event of fire, and also contains a section dealing with ventilation; B4 requires that the external walls of a building, or in this case a car park, resist the spread of fire from one building to another; F1 states that there shall be adequate ventilation; H3 covers the need to drain the runoff from a paved area; and M2 is intended to ensure (among other matters) that disabled people can obtain satisfactory access from the car park to the building.

There is also a requirement contained in Design Note 18 published by the Department for Education on the subject of access for disabled people to educational buildings, which states that reserved parking bays are required for disabled students, staff and visitors. These bays should be 2.8 m wide to accommodate vehicles used by ambulant disabled persons and 3.2 m wide if the person has to transfer into a wheelchair.

By the definition of building purpose group 7(b), which covers car parks, it is shown that the Building Regulations are intended to apply to those designed to admit and accommodate only cars, motorcycles and passenger or light goods vehicles weighing no more than 2500 kg gross. It is intended that any reference to a car park also refers to any part of a building that provides the facility to park a vehicle.

17.2 Resistance to fire and the spread of fire

The first point made in Approved Document B3, Section 11, is that the fire load of a car park is well defined and, surprisingly, not particularly high. There is some evidence to suggest that fire spread from one vehicle to another is not likely to occur and that, where the car park is well ventilated, there is a correspondingly low probability of fire spread from one storey to another. While acknowledging this low fire risk, Health Technical Memorandum 81, published by NHS Estates, states that any access from a car park to a hospital should be through a protected lobby. In addition, any vertical access from the car park to the hospital should be via a stairway or stairways serving the car park storeys and connecting with one storey only of the hospital, through a protected lobby. Apart from this particular requirement, the standards laid down in the Technical Memorandum are the same as those in the relevant Approved Documents.

Ventilation is the important factor and is dealt with in Section 17.3 below.

All materials used in the construction of a car park building should be non-combustible, except:

- any surface finish applied to the floor or roof
- any surface finish applied within any adjoining building or separated part that complies with the requirements of Approved Documents B2 and B4 (see Chapters 6 and 8)
- any fire door (which should comply with the appropriate requirements for the fire door)
- any attendant's kiosk, provided that it does not exceed 15 m² in area.

The minimum periods of fire resistance for the structural elements of a car park building depend on whether it is, or can be considered as, open-sided or not. The requirements of Approved Document B are given in Table 17.1.

A car park building can be considered to be open-sided if:

- there are no basement storeys
- each storey is naturally ventilated by permanent openings, at each level, having an aggregate area not less than 5 per cent of the floor area, of which at least half must be in opposite walls
- if the building is used for any other purpose, the car park must be separated by compartment walls and floors as appropriate.

The same considerations on whether the car park is open-sided or not

Table 17.1   Minimum periods of fire resistance for the structural elements of a car park

Level of storey	Minimum periods of fire resistance	
	Enclosed car park	*Open-sided car park*
More than 10 m below ground	90 minutes	Not applicable
Not more than 10 m below ground	60 minutes	Not applicable
Not more than 5 m above ground	30 minutes or 60 minutes if it is a compartment wall separating buildings	15 minutes, or 30 minutes if it protects a means of escape, or 60 minutes if it is a compartment wall separating buildings
Not more than 20 m above ground	60 minutes	
Not more than 30 m above ground	90 minutes	
More than 30 m above ground	90 minutes for non-structural elements; 120 minutes for structural elements	60 minutes

(*Source*: Based on Table A2 of Appendix A to Approved Document B (1992 edition))

Table 17.2   Permitted unprotected areas in car park buildings

Minimum distance (m) between the boundary and the side of:		Maximum total percentage of unprotected area
An open-sided car park	*An enclosed car park*	
Not applicable	1	4
1	2	8
2.5	5	20
5	10	40
7.5	15	60
10	20	80
12.5	25	100

(*Source*: Based on Table 16 of Approved Document B4 (1992 edition))

apply to the limit of unprotected areas permitted in the side of the building facing the boundary, as shown in Table 17.2.

17.3   Ventilation

There are two reasons for providing ventilation to a car park building. Firstly, for the benefit of persons using the car park and breathing the air within it and, secondly, for the dissipation of heat and smoke in the event of a fire. The first of these is covered in Regulation F, the second in Regulation B3 and, though the two do not seem to be related, the differences should be reconciled to ensure proper compliance with the requirement.

There is an implicit assumption in the guidance that carbon monoxide is the major pollutant source; however, a number of other variables must be considered. These include:

- fresh air ventilation rates
- emission types, rates and variation with time (most car parks are subject to periods of peak movements)
- exposure levels
- background pollution levels

17.3.1 Ventilation for the benefit of people

If the car park building is to be naturally ventilated, well-distributed permanent ventilation openings must be provided at each storey level with an aggregate area at least equal to 1/20th (or 2.5 per cent) of the floor area of which at least half should be in opposite walls. As can be seen from Section 17.2 above, this would be considered an enclosed car park.

Mechanical ventilation can be installed in addition to natural ventilation and, if so, the natural ventilation openings need only be half the area given above, i.e. 1/40th of the floor area, but the mechanical system must be capable of achieving three air changes per hour generally and a local ventilation rate of ten air changes per hour at exits and ramps – where vehicles queue with their engines running. A basement car park is difficult to ventilate naturally and, in this case, only mechanical ventilation can be fitted, giving at least six air changes per hour generally with ten in the area of exits and ramps.

An alternative approach is to provide ventilation that can be calculated so that the mean predicted level of carbon monoxide pollution, averaged over an eight-hour period, does not exceed 50 parts per million generally. In the area of exits and ramps this should not exceed 100 parts per million for any period not exceeding 15 minutes.

Further guidance can be found in the *Code of practice for ground floor, multi-storey and underground car parks* published by the Association for Petroleum and Explosives Administration or in the Chartered Institute of Building Services Engineers' Guide B, Section B2-6 and Table B2-7

17.3.2 Ventilation to dissipate heat and smoke

As already shown, the requirements for an open-sided car park are more relaxed than for one that is enclosed and the degree of ventilation determines which type it is. The requirement that, to be considered open-sided there must be openings equivalent to 5 per cent of the floor area, has been mentioned already. If the car park is enclosed, ventilation is still necessary and can be by either natural or mechanical means.

Natural ventilation in this case should take the form of permanent openings at each parking level having an aggregate area of not less than 2.5 per cent of the floor area, of which at least half must be in opposite walls. Alternatively, smoke vents may be provided at ceiling level, also with a permanent opening area totalling not less than 2.5 per cent of the floor area and arranged to provide a through draught.

Mechanical ventilation can be installed as an alternative, subject to the following conditions:

- The system must be independent of any other ventilation system.
- It must provide 6 air changes per hour for normal petrol vapour extraction and 10 air changes per hour in the event of a fire.
- The system must be designed to run in two parts, each capable of 50 per cent of the extraction rates required.
- Each part must be able to operate singly or simultaneously.
- Each part must have an independent power supply which would continue to operate in the event of the failure of the mains.
- Half of the extract points must be at high level and half at low level.
- The fans should be capable of operating at temperatures of 300 °C (400 °C in a hospital), for a minimum of 60 minutes.
- The ductwork and its fixings must be constructed of material with a melting point not less than 800 °C.

To summarize, if natural ventilation only is relied upon, the minimum ventilation area derives from Part F and is 1/20th or 5 per cent of the floor area. If mechanical ventilation is installed, it must achieve the six and ten air changes per hour required by both Part B and Part F, whether or not additional natural ventilation as set out in Part F is also provided.

17.4 Escape routes

The minimum number and width of escape routes depend on the anticipated number of people who will be required to use them. The number of occupants to be assumed in a car park is two per parking space. This total then converts into a minimum number of routes and their width by reference to Table 15.3, which is reproduced here as Table 17.3 for convenience.

The construction and formation of the escape routes, exits and fire doors from a car park follow the same rules as for any other part of the building and is detailed in Chapter 15 at 15.4.

17.5 Emergency lighting and exit signs

Chapter 15 deals in detail with the provision of emergency lighting and exit signs. All escape routes from a car park must be fitted with emergency lighting that will provide illumination for at least one hour should the electrical mains

Table 17.3 Number and widths of escape routes in a car park

Maximum number of persons	Minimum number of escape routes or exits	Minimum width (mm)
50	1*	800†
110	2	900
220	2	1100
500	2	1250
1000	3	1667
2000	4	2500
4000	5	4000
7000	6	5833
11000	7	7857
16000	8	10000
More than 16000	8 plus 1 per each 5000 persons over 16000	5 mm per person in each escape route

(*Source*: Based on Tables 4 and 5 of Approved Document B1)

* Subject to the length of the escape route, see Table 15.6.
† May be reduced to 530 mm for gangways between fixed storage racking except in public areas.

fail. Each compartment of the escape route must have at least two light fittings that will illuminate:

● escape routes to a minimum standard of 1 lux along the centreline
● exit doors
● exit and safety signs
● staircases
● fire-fighting equipment
● first-aid points

Anti-panic lighting is also required to reduce confusion if the mains fail and to aid people in finding their way to the escape routes. This should achieve a level of illumination of not less than 0.5 lux at the worst point.

17.5.1 Exit signs

These are dealt with at length in Chapter 15 at 15.4.1. As is pointed out, the prudent course to follow is to adopt the requirements of the new European Signs Directive, even though this does not correspond exactly to the requirements of the Building Regulations and BS 5499: Part 1.

The standard sign to indicate an escape route exit is as shown in Fig. 15.1 and is a pictogram of a man running, accompanied by an arrow indicating the direction of escape and a rectangle representing an open door. The sign is to be in green and white and may be lit by a light directed onto it or internally illuminated.

17.6   Access for the disabled

It must be possible for a person with any form of disability, be it a mobility difficulty involving the use of a wheelchair or crutches or an impairment of sight or hearing, to make their way easily and safely from a car park to an entrance to the building. Preferably, this should be the principal entrance but should this not be feasible, another entrance can be provided with an internal connection to the principal entrance.

Similar access must also be made available for the benefit of staff to an entrance for the exclusive use of anyone working in the building.

Preferably a 'level' (less than 1 in 20 slope) approach should be provided of a width not less than 1200 mm. If this is not possible, a ramped approach or a stepped approach can be provided; however, the latter make the building inaccessible to those in wheelchairs. The details to be followed in the setting-out and construction of ramps and steps for use by the disabled are given in Chapter 16.

17.7   Drainage

Approved Document H3 sets out the provisions to be made for the disposal of rainwater from roofs and the sizes of pipes necessary to carry the volume of water to be handled. It also states that the pipes should have enough capacity to carry the flow from paved or other hard surfaces.

The Approved Document also points out that the Building Regulations do not cover the runoff from paved surfaces but go on to specify that an allowance of 50 mm per hour rainfall intensity should be made for these surfaces.

Alternatively, the requirements of the Regulation can be met by following the recommendations contained in BS 6367: *Code of practice for drainage of roofs and paved areas* and BS 8301: *Code of practice for building drainage*. The following notes are taken from these two British Standards.

Floors of covered car parks do not, normally, need any drainage unless they are below the level of adjoining drained areas when it may be sufficient to provide a sump, with a removable cover, which facilitates pumping out.

Floors of surface car parks and vehicle standing areas should be laid to a fall of 1 in 60 minimum and drained. The levels should be arranged to avoid water accumulating alongside any buildings and channels should be provided to prevent water from running into any buildings or onto the highway.

The drainage of the paving should take account of the type of use to which the area will be put. Open parks used for nothing more than waiting vehicles

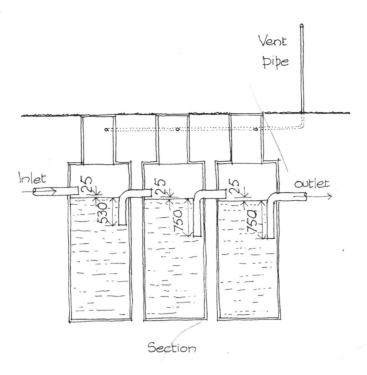

Fig. 17.1 Petrol interceptor

can have the normal rainwater drainage arrangements that would be provided for any surface water disposal. If there is the possibility that the area may be used for vehicle washing but not mechanical servicing, the gullies should be trapped and of the type with a sump in which detritus can be collected.

Any areas where vehicles are permitted to stand and either discharge petrol into a storage tank or fill their own tanks must be provided with petrol interceptor gullies. It is an offence under Section 27 of the Public Health Act to allow petroleum spirit to enter a public sewer. If this occurs there can be a build-up of explosive gases with dangerous consequences. The three-chamber petrol interceptor shown in Fig. 17.1 is the form generally constructed, but the Building Research Establishment has been testing a two-chamber design with satisfactory results.

INDEX